THE
TAO
OF STATISTICS

For information:

Sage Publications, Inc.
2455 Teller Road
Thousand Oaks, California 91320
E-mail: order@sagepub.com

Sage Publications Ltd.
1 Oliver's Yard
55 City Road
London EC1Y 1SP
United Kingdom

Sage Publications India Pvt. Ltd.
B-42, Panchsheel Enclave
Post Box 4109
New Delhi 110 017 India

Printed in the United States of America

Library of Congress Cataloging-in-Publication Data

Keller, Dana K.
  The tao of statistics: A path to understanding (with no math) / Dana
K. Keller.
        p. cm.
  Includes index.
  ISBN 978-1-4129-2473-3 (cloth: acid-free paper) — ISBN 978-1-4129-1314-0
    (pbk.: acid-free paper) 1. Statistics. I. Title.
  QA276.K253 2006
  519.5—dc22

                                              2005005273

This book is printed on acid-free paper.

       09   10   11   10 9 8 7 6 5 4

Acquiring Editor:	Lisa Cuevas Shaw
Editorial Assistant:	Karen Gia Wong
Production Editor:	Diana E. Axelsen
Copy Editor:	A. J. Sobczak
Typesetter:	C&M Digitals (P) Ltd.
Cover Designer:	Edgar Abarca

THE
TAO
OF STATISTICS
A PATH TO UNDERSTANDING (WITH NO MATH)

DANA K. KELLER
HALCYON RESEARCH, INC.

ILLUSTRATED BY
HELEN CARDIFF

SAGE Publications
Thousand Oaks ▪ London ▪ New Delhi

Contents

Acknowledgments   viii

Introduction   ix

1. The Beginning—The Question   1

2. Ambiguity—Statistics   2

3. Fodder—Data   6

4. Data—Measurement   10

5. Data Structure—Levels of Measurement   12

    5.A. Nominal   14
    5.B. Ordinal   16
    5.C. Interval   18
    5.D. Ratio   20

6. Simplifying—Groups and Clusters   22

7. Counts—Frequencies   26

8. Pictures—Graphs   28

9. Scatterings—Distributions   30

10. Bell-Shaped—The Normal Curve   34

11. Lopsidedness—Skewness   38

12. Averages—Central Tendencies 40

    12.A. Mean 42
    12.B. Median 44
    12.C. Mode 46

13. Two Types—Descriptive and Inferential 48

14. Foundations—Assumptions 52

15. Leeway—Robustness 54

16. Consistency—Reliability 56

17. Truth—Validity 60

18. Unpredictability—Randomness 64

19. Representativeness—Samples 66

20. Mistakes—Error 70

21. Real or Not—Outliers 72

22. Impediments—Confounds 74

23. Nuisances—Covariates 78

24. Background—Independent Variables 82

25. Targets—Dependent Variables 84

26. Inequality—Standard Deviations and Variance 86

27. Prove—No, Falsify 90

28. No Difference—The Null Hypothesis 92

29. Reductionism—Models 96

30. Risk—Probability 98

31. Uncertainty—*p* Values 100

32. Expectations—Chi-Square   104

33. Importance vs. Difference—
    Substantive vs. Statistical Significance   106

34. Strength—Power   108

35. Likely Range—Confidence Intervals   110

36. Association—Correlation   112

37. Predictions—Multiple Regression   114

38. Abundance—Multivariate Analysis   118

39. Differences—*t* Tests and Analysis of Variance   120

    39.A. ANOVA   124
    39.B. ANCOVA   126
    39.C. MANOVA   128
    39.D. MANCOVA   130

40. Differences That Matter—Discriminant Analysis   132

41. Both Sides Loaded—Canonical Covariance Analysis   134

42. Nesting—Hierarchical Models   136

43. Cohesion—Factor Analysis   138

44. Ordered Events—Path Analysis   142

45. Digging Deeper—Structural Equation Models   144

46. Fiddling—Modifications and New Techniques   146

47. Epilogue   148

About the Author   153

Acknowledgments

To those who lent indispensable assistance with the book:

Amy Eutsey, Diana Foley, Frank Funderburk, John Montroll, and Bill Naylor for their careful readings and advice.

Margot Kinberg for her faith, support, multiple critiques, and friendship for many years.

Rosalie Keller, my mother, who has always encouraged my scholastic endeavors.

My children: Zachary, who showed me that there are many paths to knowledge and in that way inspired the book; and Jason, who helped me to understand an artistic approach that brought the book to life.

Mostly to my lovely wife, Mary Lou, for more than I can say. I would marry her again in a heartbeat.

About the Illustrator

Helen Cardiff is an artist and illustrator living on the Eastern Shore of the Chesapeake Bay in Maryland. Her work is prominently displayed during yearly civic events as well as in galleries and private collections in the United States, Canada, and Europe. The fusion of her Taoist and artistic trainings fosters fresh insights into the book's topics. She can be reached at hcardiff@bluecrab.org.

Introduction

For most people, the concept of statistics begins as a shadowy mathematical nightmare, followed by a mandatory course or two in college, and is considered a weakness forevermore. Why? The answer is incredibly simple: The way statistics is taught is not the way most people learn. Most of us learn easily through impressions and experiences, the qualitative side of life, not through numbers and equations, the quantitative side. This book shows that the fundamental concepts of statistics can be learned through impressions and experiences, too. Simply put, understanding the meaning and worth of statistics does not require the ability to calculate them.

Statistics are ways of understanding more about questions and issues that interest us. *The Tao of Statistics* is a journey down a path that leads to an intriguing view of the world. The statistical view of the world is of a place where knowledge is neither certain nor random. Statistical portraits are painted in pastel rather than in primary colors. Extremes are understood for the information they contain but are done so in a context of centrality. *The Tao of Statistics* lays a path to this understanding of the world. The path leads to a view of the subtle patterns in life that were invisible before.

The world has changed. Only a few decades ago, a relative handful of people could manage society's needs for calculating and interpreting statistics. Now, paradoxically, an even smaller group needs to know how to calculate statistics, because computers have taken over much of the calculation. This may be both a blessing and a curse. Vastly more of us, however, need to understand what statistics are telling us about our work, our communities, our children's test scores, and even our play.

This larger group tends to have a more qualitative learning style than statisticians of years ago, yet this difference (i.e., "the rest of us") has been largely overlooked by writers of textbooks on the topic. Regardless of the prestige of the school or the academic domain, statistics almost universally is taught through equations, hour after brain-breaking hour. What is the point of this traditional chore? Apparently, it is to inculcate an intuitive understanding of statistics that will follow from it. Is there evidence to support this expected outcome? That is, do we live in a nation with a collective intuitive understanding of statistics? I do not think so. More likely, we live in a nation that cringes at the term *statistics* and is skeptical of their use. It does not have to be this way.

This book introduces readers to the commonly encountered terms and techniques in statistics through verse, graphical illustrations, and accompanying text. No equations are used. The result is an understanding of each concept that is created without having to do math.

This journey is about living a series of experiences, their impressions, and sometimes even their emotions. These experiences lead to a deeper understanding of the world of statistics as we follow the paths of two research-oriented professionals with different careers as they experience the statistical concepts presented herein. Try to experience the concepts with them. Stroll down the path together. Before beginning, though, a few thoughts on statistics themselves might be helpful.

- Statistics are filters on how we see the world. They focus our vision, and they help us to see through the fog. In doing so, they also prevent us from seeing some of what else is there. Stay aware of what is being filtered out, too.
- You do not need to know how to calculate statistics to understand what they are telling you.
- Averages do not exist for most things. They are only ideas. A truly average person does not exist. The idea of an average is what is useful.
- The rest of statistics is no more real than averages. Regardless of how technical a statistic sounds, it is still only an idea that can be grasped by all.

- The world of statistics has become technically quite complex. When someone produces a statistic that you have never heard of or seen before, simply ask what it does and for an example that demonstrates its usefulness.
- Ideas lead to understanding. Experiencing the ideas of statistics motivates us to develop and deepen our understanding of them.

This journey is best experienced in comfortable clothing, in a favorite chair, and in a quiet space.

Take your time.

Breathe deeply and slowly.

Relax.

Let the world pass you by.

For each section, try to open yourself up to the impressions from the verse and from the graphic. Next, try to put yourself in the shoes of each of the two professionals we follow, as they use statistics in their otherwise specialized jobs. Then, reread the verse, pausing for a few moments, and look again at the illustration to deepen your understanding of the concept and help you to anchor it in your long-term memory.

Finally, be ready for a noticeable change. Life never looks quite the same again after being seen through statistical eyes. It is more predictable but less deterministic. Mostly, one acquires a growing sense of wonderment at the patterns and relationships that emerge from the chaos.

If you enjoy this journey, you might also value the result of learning the math behind statistics. With the mathematics comes a much richer appreciation of the statistical view of life. The precision with which it approaches ambiguity is simultaneously humbling and amusing.

1. The Beginning—The Question

The world is hazy

No clean lines or sharp points

Focus the camera

The world of statistics starts with a question. Many types of statistics exist because questions can be of almost any type, can be about almost anything, and can take place under quite varied conditions. This variety calls for an almost equal number of statistical approaches to answers.

Statistics need data. Data come from measures. Good answers to important questions require good data from relevant measures. Good measures are specific and, through that specificity, are informative. If the selected measures do not directly address the questions, the statistics can be meaningless because they will have lost their context.

Statistics become important when they are about things that people consider important. In this book, we will follow two professionals with different careers as they use statistical concepts to answer questions that are relevant to their own lines of work. The first is a high school principal with questions about differences in quality and other characteristics of the various classes taught at his school. The second is a director of public health with questions about how well the residents of her state are meeting national guidelines for public health issues, such as immunizations, especially for individuals receiving public medical assistance.

The choice of careers and the specific concerns of the related fields really do not matter to the interpretation of statistics. *What matters is that appropriate data are used with appropriately selected statistical techniques in contexts that address important questions in interpretable ways.* In fact, we will see that the data and statistical issues facing both of our subjects are far more similar than they are different, although their contents greatly differ. The process of *doing* statistics in research projects changes very little across a wide variety of questions and fields of interest.

The world of statistics starts with a question, not with data.

2. Ambiguity—Statistics

Not enough to know

Just enough for a better guess

Statistics are born

S tatistics help to tame ambiguity by quantifying it (a point we will revisit a few times). In the world of statistics, if there is no ambiguity and no need to guess, we use *population parameters*. Where there is ambiguity, we use *sample statistics*. These terms have been shortened over the years to *parameters* and *statistics*. Statistics, then, are ways to make educated guesses. They might do so with a remarkable flair for Greek letters and long equations, yet they are guesses nonetheless.

Samples have *sampling error*, again by definition. All statistics start from samples and have various amounts and kinds of sampling error. Samples can even be *samples in time* but must be samples of something bigger if you expect to produce a statistic. Because it is generally too expensive to measure everyone or everything representing a situation of interest, statistics are used throughout academic, professional, and everyday life.

When they are properly phrased, statistical results are hard to disprove because they do not contain absolutist language. This built-in ambiguity can be frustrating to those who want a strict yes-or-no answer, especially after they have waited until the data are collected and analyzed, sometimes at great cost. Unfortunately for these individuals, statistics are not meant to suit the convenience of the moment. Only so much can be properly inferred from a system of educated guesses, regardless of how carefully the guesses are made.

It would not be much of an exaggeration to say that the world is run by statistics, or at least by people who get statistical information upon which to base their decisions. With the communications industry almost omnipresent in day-to-day life, people's interest in and need to understand statistics has unnoticeably mushroomed into a dominant feature of modern life. Why? Because statistics are used to answer people's questions, and the answers reported (through TV, radio, newspapers, etc.) use statistics as evidence. Remember, research needs samples, and samples generate statistics.

In our day-to-day lives, we use statistics without even knowing it. Those of us who own and drive a car guess whether we can make it to the next gas station based on what we know of the road conditions, typical gas mileage (a statistical issue, to be sure), and the consequence of being wrong should we run out of gas while on the way to where we are planning to go. Non-drivers make equally statistical estimations and decisions, such as how much cash to keep on hand for a given weekend's planned activities.

In our first encounter with the high school principal, we see that he has data that are capable of addressing a wide variety of relevant academic and social questions. His questions are mostly about current academic achievement, but some are more future oriented. He wants to use his data to support budgeting and expansion activities as well as to identify current problems for more immediate attention.

The principal considers all of his data to be samples, even when a casual observer might suggest otherwise. He wants to generalize to other classes and years. His 20 years at the school have shown him that changes in the makeup of each student class occur very slowly and subtly from year to year. For him, statistics are safer than having to take a harder stance. He likes to fall back on statistics being a science of quantifying ambiguity, and that means that his answers will be somewhat ambiguous, too.

The director of public health will have access to very large and mostly representative samples to address her questions. Although her questions will be answered through her statewide electronic database, people are missing, for a variety of reasons, throughout the data. Yet her data are more representative than most and are likely to be consistent from year to year, even given any unknown sources of bias. Given the importance of year-to-year comparisons and the need to be sensitive to people moving in and out of the medical assistance program over time, she is quite pleased with the representativeness, completeness (the relative lack of missing data), and comprehensiveness (the availability of measures to address important characteristics, for her questions) of her data. For her, ambiguity is part of what gives her the luxury of testing different hypotheses for the potential impact of changes in public health policy.

This last aspect of ambiguity holds one of the keys to the path of statistical knowledge. The ambiguity in the system means that no known solution fits perfectly. Ah, the challenge! Think of it! What is the best solution? Is the best solution the one with the smallest errors, or is it the most parsimonious one? What data are available?

How good are they? Judgments fly. Decisions are made. The challenge in defining a statistical solution often is not to be the most correct but, instead, the least wrong! How? Through a sharp question that cuts straight through ambiguity associated both with the statistical and methodological approaches and with the data themselves.

The overarching message about statistics is that they are uncertain. Treated that way, statistics become more of an intellectual challenge, with less attachment and much less certainty. So, continue to relax and marvel at some close-up views of the foundation of statistics. Look at where the cracks are. Realize what those cracks could mean for edifices built on top. Smile, as together we experience more about this view of the world.

3. Fodder—Data

Observe

Record

More

D ata are what we hear, see, smell, taste, touch, and more. Data can even be what we sense. Data can represent anything and everything that we can discriminate well enough to distinguish from something else. In short, if it can be perceived, it can be coded and used as data.

Data are the fodder of measurement, the backbone of statistics. Through a context, data become transformed into information. That context is a fusion of substantive knowledge of a topic with a methodological approach to gathering the data and the statistics used to derive meaning. A large part of the misuse of statistics is a nonreflective, uncritical crunching of numbers (i.e., data) to generate other, somewhat context-free, numbers. These uncritically examined results are then granted trusted status based on unfounded validity (discussed later in some detail). The result could be a poor decision or an ineffective policy; yet, the statistics eventually are blamed. To become useful information, good data need to be placed in relevant contexts, with clear understandings of the strengths and weaknesses of the statistics and results.

This relevant context is the frame of reference from which relative meaning is derived. To know whether something is big or small, there needs to be a question of *compared with what?* A blue whale is small compared with the planet. An ant is huge compared with atoms. This same issue of needing a frame of reference, a comparison point, is important to most types of knowledge that might be acquired through statistics. Several types of frames of reference exist in statistics, as we will see.

One brief side note on the word *data: Data* is a plural word. Until very recently, the only proper grammatical use was as a plural noun, such as geese. Correctly, then, data *are* transformed through a context into information. A single piece of data is called a *datum*. With all that said, a recent English dictionary has recognized the common use of *data* as a singular noun and grants that use as a secondary preference.

Modern databases can contain dozens of gigabytes of information—an amount that is truly staggering to consider. High-speed office computers can need hours just to run through the data once. Census data are now available across the Internet. From course catalogs to recent golf scores to real-time stock prices, data surround us as oceans surround fish. Data are everywhere and generally too common even to notice.

Here is where the tao of statistics starts to take shape. Curiosity births questions that create the need for data that come from measures that people design to create meaning. We open our eyes with questions and perceive contextually rich data as probabilistic answers. Depending on how we ask our questions, how we look for and process the data, and how we place results in a meaningful context, the tone and the texture of the results will differ. Even with the most evenhanded intentions, unconscious biases can creep into even the best of research designs and processes. We will touch on this point several times in later chapters.

The high school principal has student records in an electronic form, meaning that his data collection will be inexpensive. Having electronic student records also means that the principal has access to a wide variety of data for his students. Throughout the years, the school system has collected demographic data on its students. The principal also has the funds for conducting a survey on his essentially captive audience. Although he is not new as a principal, the extent of the electronic data available has him a bit intimidated. When he used to have to get the data from students' physical files in the office, his "research" questions were quite modest and constrained. Now that he can get hundreds of times the amount of information with only a few mouse clicks, he is somewhat more reflective, less impulsive, less likely to "just run the data" than he had thought that he would be.

The director of public health has all of the state's Medicaid information available to her electronically. She also is authorized to conduct a single, limited survey if it can be seamlessly appended to one that is currently required by the state. She has less information on each person than does the high school principal, but the information she has is for a much larger number of people. When she accesses the data warehouse, she always pulls highly detailed data (i.e., *disaggregated*). She knows that she can always collapse (i.e., *aggregate*) it later, but not the reverse.

Having data for large numbers of people and access to computers allows her to address important public health questions that would have gone unanswered not many years ago. Just as the principal has access to far more information than he used to have, the director of public health had that increase in access several years earlier. She is used to the amount and has started to understand the data's strengths and limitations.

Both the principal and the director of public health face the issue of data privacy for the individuals for whom they have data. Well-established protocols exist for the proper handling of this issue. Remember, data privacy has ethical and legal standing. Expected processes and procedures exist for research involving people and their data. Keep the importance of this issue in mind when using or when reporting human subjects' data. Ignorance is not a valid excuse.

4. Data—Measurement

Perceivable

Describable

Scores

If you can perceive it, you can measure it. A *measurement* is an assigned value for a single characteristic. The way a characteristic is captured and, therefore, the way its data should be interpreted determine the *measure* being used to address the question at hand. Some measures are more accurate than others. Perfect measurement exists only in fantasy; we do the best we can.

Good measurement not only is sufficiently accurate but also places its objects into mutually exclusive categories or scores (or "codes"). Some measures divide people into categories, such as gender. Other measures are more abstract continua, such as perception scales that ask the extent to which a respondent agrees with a statement. Regardless of the type of measurement, sufficient accuracy and mutual exclusivity are needed. The rest of measurement is an extension of those simple concepts.

Data . . . the who, what, where, when, why, and how. Put the pieces together like a jigsaw puzzle, and voilà, you have meaningful information from what was a pile of otherwise useless facts. Be careful: In statistics, as in a jigsaw puzzle, cutting corners and force-fitting pieces can result in a very misleading picture. These shortcuts are sometimes difficult to notice and even more difficult to resolve.

Along with grades, the principal's school keeps information on standardized test scores, disciplinary actions, health records, extracurricular activities (e.g., clubs), and sporting achievements. Depending on state and federal laws, the principal will have varying levels of access to student records.

The director of public health has access to all of the public health and some other state databases. Again, her access is legally limited because the data are about health issues, such as immunizations and outbreaks of certain reported diseases. State and federal laws are quite strict on the access to and use of these types of data.

5. Data Structure—Levels of Measurement

What can be built?

Ask the ground

Turn over rocks

Dig in the dirt

T he grounding for statistics is the level of measurement of the data. Some statistics are appropriate for some levels of measurement; others are not. This is an area where one needs to understand the deeper structure of the data to know which statistics would be meaningful. For example, the data's level of measurement limits the choice of the most often-used statistic—the *average*, what statisticians call the *central tendency*. There are three common choices of averages: the *mean*, *median*, and *mode* (with somewhat esoteric versions within each). These different types of averages are not equally appropriate for data at different levels of measurement. Specific levels of measurement and the impact of each on the choice of statistics will be discussed soon.

The topic sounds complicated, but is not. Once you understand how data differ according to their level of measurement, you will quickly grasp which statistics are appropriate for a given set of conditions. Fortunately, many statistical techniques have options that can account for the various levels of measurement.

Through the questions being asked by the high school principal and the director of public health, we will encounter four of levels of measurement (i.e., *nominal*, *ordinal*, *interval*, and *ratio*, to be explained next) in their various data sets or from potential survey responses. They, and we, will accommodate these levels of measurement as we progress through this book.

Important to getting the statistics correct is the recognition and accommodation of each variable's level of measurement. Even researchers with decades of experience occasionally will be embarrassed by having used a statistic in a way that was inconsistent with the level of measurement requirements of that statistic. Though an issue with level of measurement rarely is a knockout punch, these issues tend to be varyingly important limitations on the confidence that researchers (should) have in their results.

5.A. Nominal

Nominal says Different

No more does it claim

Others shouldn't either

The *nominal* level of measurement is about categories. Some statisticians refer to it as the *categorical* level of measurement. The categories have characteristics that differ but are not quantified as to the amount of the difference. For example, political party, religious affiliation, gender, and so forth can be recorded, grouped, and counted. Yet we do not say, for example, that one religion is more of a religion than another.

Under certain conditions, the most typical type of average, the mean (i.e., arithmetic average), is appropriate for nominal data. That is, the variable has only two possible responses, and talking about the percentage that corresponds to one of those responses makes sense. With gender coded 0 for female and 1 for male, it would make sense to use the mean to say that a group is 60% male.

Variables coded and interpreted as we have just seen find use in a variety of statistical techniques requiring at least interval levels of measurement (a discussion about interval level data is coming shortly). Some nominal data, therefore, can be quite useful in answering a surprisingly broad range of questions.

Both the high school principal and the director of public health have nominal data. The high school principal has data on gender, school club membership, sports participation, and scholastic topics for each student. Some aspects of these measures could be coded as nominal, such as variables for the names of extracurricular activities (e.g., yearbook).

The director of public health has access to a host of demographic data that are nominal, such as ethnicity and zip code. Generally, nominal data are summarized in tables or cross-tabulations of two characteristics, such as sports participation by gender or immunization rates by age or age grouping. Nominal data also delineate many of the groups of interest to research.

5.B. Ordinal

With distances unsure

Blindly even steps

Arrive at cracks

Ordinal measurement is common for opinion polls. We can distinguish between levels of agreement but cannot be sure that the psychological distance between pairs of adjoining response choices are equivalent. For example, the psychological distance between "strongly disagree" and "moderately disagree" might not be the same as the distance between "neutral" and "moderately agree." In these cases, an arithmetic average (the mean) might not yield an interpretable answer.

The high school principal has ordinal scales from some student surveys that he has already conducted, and of which he might generate more. Although the case could be made that course grades really are ordinal, they have been and continue to be used as interval (the next topic) since their creation. The debate is whether the difference in knowledge of a topic between two students scoring, say, 20 and 60 points on a test is the same as that between students scoring 60 points and 100 points.

Along with actual medical data, the director of public health has results for perception surveys on the services received by the state's medical assistance recipients. She also has another survey to be implemented fairly soon, a state requirement of her department. Most of her medical data, however, are either nominal or ratio, at least in how they are handled.

For statistics appropriate to ordinal data, both the high school principal and the director of public health will use frequency counts for the responses to each of their surveys' items and a form of chi-square (described a bit later) for statistical significance tests. They both will use medians and modes (also discussed later) to describe these central tendencies. Recognizing ordinal data for what they are can save many later headaches. Statisticians using ordinal data with statistics requiring interval data sometimes pay a harsh price in terms of their reputation.

5.C. Interval

Interval is regular

Same steps, no cracks

Yet zero is not none

Interval data have evenly spaced steps but no true zero. Course grades could be an example, where a zero score on a math test does not mean a complete lack of mathematics knowledge. A zero on a math test means that the student did not arrive at a single correct answer for the sample of possible relevant mathematics questions on the test, but the test has no way of capturing whether the student has *no* knowledge of the assessed topic. The zero is a measurement convenience.

Many statistics require interval levels of measurement (or could use *ratio,* discussed next) to yield valid results. Topics from grading differences in sections of the same course to the predicted flu infection rates for next year generally require this level of data. At a minimum, some reflection is appropriate when determining which statistics will be used with data.

For the high school principal, most student achievement measurements are used as though they were at an interval level of measurement, as discussed earlier. The very fact that the debate continues, more than a century since its inception, is testimony to the resiliency (what statisticians call "robustness") of the mean to minor violations of its required level of measurement.

Many examples of the director of public health's data are dichotomous (i.e., only two possible responses). For example, immunizations are coded for people in one of two ways, either yes (1) or no (0). These types of data generally can be used in statistical techniques that assume interval levels of measurement.

Examples of true interval data are somewhat rare. The most common are Fahrenheit and Celsius scales to measure temperature. In the end, the interval level of measurement is important to the proper selection, use, and interpretation of statistical methods, but it has few true examples in daily practice.

5.D. Ratio

The rare ruler

The flexible measure

Precious property

A *ratio* level of measurement scale has a true zero and is the trophy of data types. Weight and height are examples. We can say that half of 100 pounds is 50 pounds, and twice 6 feet is 12 feet. In other words, we can form interpretable ratios. These types of data are almost carefree in their use with regard to their level of measurement (assumptions on their distributions is another story, which will be told shortly).

The high school maintains basic health information in the nurse's office on such things as height, weight, and inoculations, but the high school principal would likely need a good reason to be granted access to many of these data. The principal does have, however, somewhat unlimited access to absentee and tardiness information. These variables are at a ratio level of measurement. Depending on how the data are coded and used, though, they could be at any of the levels of measurement. Recoding (assigning new codes after the fact) can further complicate understanding the data's true level of measurement.

The director of public health has electronic medical information for all Medicaid recipients in her state, although restrictions on the data's use are quite stringent. Nonetheless, for a variety of purposes, she might have access to any of the types of information, including measures that have true zeros. For example, when looking at the degree of compliance with national adult immunization guidelines, she knows the number of different immunizations appropriate to adults and the number of immunizations delivered. Knowing who received which immunizations and counting them for each adult, she can form ratio scales for immunization counts and for rates of immunization compliance. The more her data are measured on interval and ratio scales, the larger the variety of statistical techniques that will be available.

6. Simplifying—Groups and Clusters

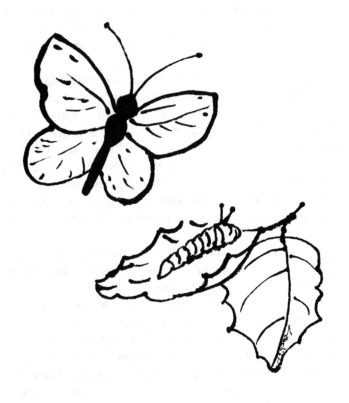

You and I are much alike

Over there, they are different

I will be with those like me

For now

G roup differences are the cornerstone for much of the social research done today. The reason is that grouping is a convenient, logical, and valid way to reduce the complexity of data. *Groups* are created from ideas that people have about characteristics or conditions that separate interesting parts of the data. Gender and race/ethnicity are traditional examples and are used more often than they are reflectively reviewed for their relevance to answering the questions posed.

Clusters are groups that are created by sophisticated mathematics when researchers do not know where, or how, to separate their groups. To form clusters, statistical software needs variables, which someone must choose and code. To that extent, researchers need a fair idea of the characteristics that would distinguish between groups, more than a random guess. The more researchers know how groups were formed in the data, the better they can utilize group information to address interesting questions of the data. If I wanted to know the characteristics (i.e., through clustering) to predict whether a diabetic will have a biennial eye examination, I would start with as many demographic and health care access variables as I could find. I would not look for hat size or color preference information.

Both groups and clusters are used to understand inequalities in life. Also important, they can highlight similarities. Frequently, created clusters are used as though they were naturally occurring groups. With empirical evidence forming the foundation for a given cluster, is prior knowledge and recognition of the commonalities among members of that cluster required to consider members a group? Regardless of the answer to that question, retaining a semantic distinction between groups and clusters increases the specificity of results and adds context to the discussion.

The distinction between using groups versus clusters is whether a strong hypothesis on the source of group differences exists prior to the start of the research. Researchers have found that, lacking a strong hypothesis, letting the data speak for themselves (i.e., clustering) can provide interesting insights. Such insights, though, still need a substantive context.

Analysis of voting patterns by age bracket is a familiar use of groups. News reporters use groups in this way to show differences in voting preferences for young, middle-aged, and older adults. Clusters, on the other hand, might be formed to find common characteristics among members of a particular voting bloc, say those who voted for an independent candidate. Often, one finds that traditional groupings are used (e.g., gender) when such groupings are not related to the question of interest, a situation that should make one pause.

The high school principal has all manner of information on groups: year in school, sports team membership, demographics, courses, and separate sections for the same courses, to name a few. Differences in students' grades across these groups might suggest some important inequalities in his school. If vastly different grades, for example, were being awarded for similar work by similar students, an intervention with the teachers might be warranted. The statistics would not suggest which teacher graded too high or which too low. Those are value judgments. As we have seen, statistics do not make value judgments. They are remarkably evenhanded, even though statisticians might not always be due to unconscious biases (more on this later).

The director of public health also has access to data that easily can be formed into groups. Nonetheless, she might want to create clusters to address questions about immunization patterns. The literature is not as clear as she would like for purposes of instituting policy changes aimed at improving adult immunization rates. She needs to understand local differences in patterns for people with higher immunization rates compared with those who have lower rates. Using cluster analysis to form groups is a viable method for using her data to understand important differences in these rates. Looking at the key variables selected by the computer would help her to understand the differences between clusters that might be used to leverage her resources with the lower-scoring clusters.

Fortunately, the director has very large samples with which she can test large numbers of combinations of variables in computer runs that would have been impossible only a few years ago. In minutes, she can test the most likely 30 or 40 variables. Once she sees the major contributing characteristics for cluster membership, she can look at the socioeconomic and cultural structures of the area, then work with local expert interventionists in support of public health care policy.

A few words of caution are in order for using groups and clusters. Groups and clusters simplify analyses by reinforcing the notion of differences between people or events. The potential for stereotyping, intentional or not, is high. Appeals to a higher authority, such as the funding source, are not sufficient justification for unreflectively highlighting differences that are not relevant to the question or are wrong by sometimes subtle inference. No method beyond honest reflection exists that will ensure the ethical use and labeling of groups or clusters.

To walk the path of statistical knowledge is to remain aware of unintentional damage that unreflective analyses might cause. These are the latent functions (the unintended consequences, in policy and program evaluation terms) of policy or reporting, be it public or private, large or small. Either formally or not, statistics are often used as evidence to support one perspective at the expense of another. Think about what those perspectives might be and whether you would want to be associated with them.

7. Counts—Frequencies

How many here?

How many there?

How many everywhere?

*F*requencies are counts. What they count and how the counts are split into groups mainly depend on the question asked but frequently depend to some extent on the data (i.e., their levels of measurement and distributions). Frequencies are meant to convey a sense of both absolute and comparative magnitudes. Both counts and percentages are generally used to present this information. Usually, we do not see more than about 6 or 8 groups in a set of data, except when they naturally fall into other categories, such as the 50 states (plus Washington, D.C.). Even then, the categories are often regrouped into a smaller number of categories (e.g., New England, Midwest).

The high school principal wonders about the effect on his students' grades from high commitments to extracurricular activities. He decides that a good measure of this commitment would be the number of such activities in which each student participates. He produces a table that shows both the number and the percentage of students who do not participate in any activities, those who participate in one activity, two activities, and so on. The student with the highest count participates in seven extracurricular activities in a single year. Although almost half of the students do not participate in any activities (46%), of those who do participate, almost two thirds (63%) participate in three or more activities. This information supports policies that assist students who take advantage of these types of opportunities not available elsewhere in their communities.

The director of public health wants to know how often local emergency rooms are used as though they were a physician's office. She wants to know the extent to which expensive emergency care is used in place of less expensive primary care and how that use is changing over time. Her findings are mixed. Fewer people are misusing the emergency room in this way, but those who are do so more frequently than in years past.

8. Pictures—Graphs

Closer, maybe squint

You give it shape

Meaning appears

G raphs are important for many of the same reasons that distributions are important, plus one more that is critical: Graphs are pictures. Most people are visual learners, and there is real advantage to capitalizing on that fact. Vision is packed with information about our environment. Used in reports, graphs are concise conveyances for much information. Graphs help to inform statisticians on the likelihood that their assumptions (such as heteroskedasticity and linear versus curvilinear relationships, but these need not concern us here) are met.

Statistics in a vacuum are about as useful as a burnt match. Graphs communicate a sense of context, bringing meaning to the statistics. Yet, like a magician's hands, graphs can mislead, either by intent or by carelessness.

When preparing graphs, simpler often is better. "Busy," cluttered graphs distract from the central point being delivered. Graphical representations are opportunities for elegance and style in presenting data and for the delivery of relevant information. Try not to waste these moments when data can come alive.

The high school principal is not very skilled at creating graphs, and his self-confidence is not strong in this area. He will be getting some help from his secretary, who is quite accomplished with graphical and presentation software. She takes pride in presentations that look professionally developed.

The director of public health is quite expert at generating graphs as well as in interpreting what she sees in them. She often uses graphs and enjoys adding impact by using gradations of color and texture. She is also comfortable with explaining graphs to large audiences. Over the years, she has collected a rather large directory of presentations. Now, she can quickly find slides that she has used previously and that can be altered easily to fit her current needs. With the pool of ready material, she finds that she can concentrate on her message without being as distracted by having to create the presentation.

9. Scatterings—Distributions

Diversity is information

The foolish fight it

The astute understand it

The wise swim in it

To address the questions of who, what, where, when, why, and how, statisticians need to know how data are spread and shaped. The *normal curve* (covered in Chapter 10) is arguably the most important shape (i.e., *distribution*) in statistics. Knowing about distributions is part of matching the correct statistical technique with the question being asked. Further complicating the issue of distributions is that statistics themselves have their own distributions. The term *distribution* sounds deceptively simple for what it entails. Fortunately, many statistics that rely on the normal curve are sufficiently robust even to moderately large violations of their assumptions; that is, the results can be fairly close to what would be found if the distributions were closer to the so-called normal. Other distributions start to approximate the normal curve only with sufficiently large samples and under some other conditions.

One such distribution is the *binomial* distribution, where the basic unit forming the distribution can take only two values, such as for coin flips. Any event or characteristic with only two values or outcomes, where every person or data set member has only one of them, forms a binomial distribution. With a large enough sample and under the right conditions, the binomial distribution comes close enough to mimicking the normal distribution that many statisticians treat it as though it were a normal distribution right from the start. When all is said and done, the normal distribution is at the heart of most statistics seen by the general public.

Distributions are sources of information about the inequality of characteristics. Metrics for this inequality are units such as standard deviations (discussed later in the book, specifically in Chapter 26). Knowing about inequalities is the first step in overcoming them. Statistics in a context become tools of policy, levers for perspectives. Think of this point often. When researching inequality, one is often seen as passing judgment on policy and practice. That is because policy and practice normally claim to be focused on reducing inequalities. So, reporting inequalities could be interpreted as stating that the relevant policy is a failure because there is still a need for it. In fact, the opposite could be true. The policy could be filling a need so successfully that more of those with that need are coming forward to be served by the policy.

The world of statistics is very much a world of distributions. Measures have distributions. Statistical tests have distributions. Statistical results have distributions. Even distributions have distributions. Although statisticians need to keep them all sorted out, most people do not need to be concerned at that level of detail. Yet even a brief swim in this sea of distributions should be enough to make one a bit wary about whether any type of certainty exists in statistical results.

Both the high school principal and the director of public health understand the importance of distributions. Both have decided to use statistical techniques that are robust to at least minor violations of the assumptions of normality. They both have dichotomous data for gender. These data form binary distributions because they can take only two values. Additionally, both researchers know that many of their other measures should correspond at least roughly to the normal curve by the nature of the measures alone, such as height and weight.

Nonetheless, distributions are an area where one needs to be fairly certain about what one has before running most complex statistics. Most experienced statisticians graph their data, along with generating some descriptive information, to make an informed decision about whether they have what they think they have. This issue takes judgment, and that judgment often comes from learning from mistakes. Part of the art of statistics is understanding the subtle relationships between the theoretical distributional underpinnings for a chosen statistic and the distribution presented by the actual data.

After some thought, both the principal and the director of public health believe that in addition to their dichotomous data, they also have nominal and ordinal data that do not sufficiently correspond to a normal curve. These data need to be handled differently. Recall that the level of measurement determines which statistics are appropriate for data. Nominal and ordinal data do not correspond to the normal curve (discussed in more detail in the next chapter) except under special circumstances of dichotomous data, as discussed earlier. Statistics that capitalize on the normal curve are called *parametric*, and those that do not (and are appropriate for nominal and ordinal data) are called *nonparametric*.

Visually, some distributions are easier to recognize than are others. When graphed with the appropriate rounding and scaling, more distributions become visually recognizable. With practice, some statisticians learn to estimate the correlation between two variables with surprising accuracy by simply looking at a scatter plot of their data. Time after time, when results seem just a bit odd, the researcher graphs the data, and a distribution is found to be other than was thought. Perhaps sometime in the future, statisticians will always graph their data before running models on them. It does not always help, but it cannot hurt.

A final point on distributions is much like the final point on groups and clusters. Researchers need to reflect on the plausible consequences of showing distributional differences by groups or clusters. Apparent differences in the sample might not be reflective of the truth in the population; therefore, presenting distributional differences that some groups or clusters might find offensive risks much, on little sound evidence. Relevant statistics inform a context. Be as sure as you can that you want to be a part of that context.

10. Bell-Shaped—The Normal Curve

Cardiff

School bells ring
Some melodious, some not
The faces of the new world

The *bell curve* and the *normal curve* are two names for the same thing. Originally, it was called the normal curve of errors. This term comes from some of the earliest work in statistics, which focused on predicting a son's adult height from the height of his father. The errors from these predictions formed a bell-shaped curve. Other characteristics were then predicted. The errors from those predictions also formed bell-shaped curves. Remarkably, many common characteristics did the same thing. That bell-shaped curve was then named the normal curve of errors. Since that time, statisticians have found that many common traits are distributed the same way. The mathematical properties of the normal curve empower a much-used branch of statistics, called *parametric statistics* because its properties conform to the parameters, or shaping characteristics, of a normal curve. You saw the term *parameter* used in descriptions of populations. That is because populations also have shaping characteristics. The two uses of the term are related, although they function somewhat differently.

The normal curve is pivotal in much of statistics. Converting data to their normal curve equivalent values allows an incredibly diverse range of statistical techniques to be applied. This portion of statistics is involved with z scores, the normal curve equivalent values. These values are measured in *standard deviation units*, covered later in the book (specifically in Chapter 26).

The normal distribution of errors themselves is also central to modern statistics. This distribution of errors underlies *confidence intervals* (covered later, specifically in Chapter 35) and other quantifications of the amount of ambiguity in statistical results. Without the normal curve, statistics would be a faint shadow of what it is today.

As mentioned before, under certain conditions, binomial distributions approximate the normal curve. Here is how. (People actually enjoy the following exercise.) Let us say that there are 20 people in a room. Form 10 pairs of people. Have one of each pair flip a coin 10 times. Have the other count the number of heads. After all teams are done, plot the number of heads that each of the 10 teams got in its 10 flips, from 0 to 10.

Here is how. Draw a horizontal line near the bottom of a wall chart. Mark off the line with a scale that ranges from 0 to 10. Have the teams report how many heads out of their 10 flips they had, one team at a time. When a number is called, place an x over that number on the line. Where more than one team has the same outcome, simply plot a second x above the first. Tell the teams to switch flippers and counters and do it again. After only a few rounds, the normal curve starts to take shape on the chart, right before people's eyes. The high point of the curve will be at about 5 heads, with the distribution falling off in both directions toward 0 heads and toward 10 heads. Some underlying rules for the sample size and other issues also come into play, but binomial distributions often are treated as normal distributions anyway. After doing this demonstration a few times, you can understand why. A reasonably normal distribution forms every time.

Both the high school principal and the director of public health have access to statisticians. The high school principal has access through his school board, and the director of public health through her department. These services are not inexpensive, though, and are limited by budgetary allotments. Both the principal and the director decided to conserve their resources and use a statistician only for their more complicated analyses and their more controversial issues.

The director of public health differs from the high school principal in her use of a statistician in one important way. She knows that she is working with large databases and that many aspects of her research and data require experience to arrive at valid results. Therefore, she decides to consult with her statistician early in the process, so that she does not waste resources as her statistics start to unfold. With large data sets, having to re-run analyses is time-consuming and usually wasteful of paper. Because writing reports tends to fill whatever time is allotted, the director would rather spend more of that time thinking through a discussion of her results than in re-running the data to get them.

The principal's view is that his distributions are not sufficiently complicated and his work does not have a high enough profile to justify a statistician's time in the early stages. Moreover, much of his work involves trending the same measures each year. He uses these measures for parent and school board meetings. Because he has had few problems in the past, he does not see a need to be any more cautious now. Although he is probably correct, errors at this point could be quite costly later.

The reason for the later cost from early mistakes is that distributions form the backdrop for the probabilities that are the results of statistics. These probabilities (discussed in various contexts throughout the rest of the book) are the quantification of ambiguity and inequality in the situation represented by the data. Out of distributions, statistical results arise, so we need at least a passing acquaintance with how our statistics and our results are shaped by the constraints and strengthened by the support of the normal curve.

11. Lopsidedness—Skewness

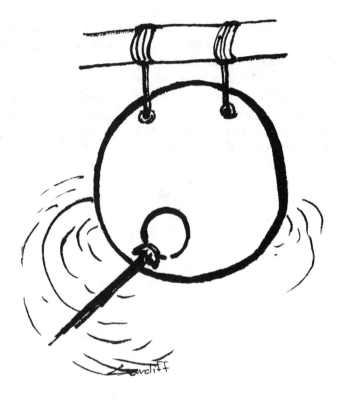

Errors unbalanced

Strength erodes

Answers suffer

S ymmetric distributions are like paper dolls: They look like folded paper that was cut and then opened to show a balanced shape on both sides. *Skewed* distributions are lopsided, off-balance. Parametric statistics depend on a balanced, normal (bell) shape. When a normal curve looks pulled to one side, the distribution is said to be skewed to that side. Income distributions, for example, are positively skewed because relatively few people earn many times what typical people earn. Small amounts of skew are no problem for most statistics. Large amounts can bias results.

Methods exist to handle skewed distributions. Statisticians have developed ways to handle most troublesome statistical issues. The problem is that the statistics get very complicated very quickly. Fortunately, relatively few people need to know anything more about the issue than to ask whether it has been handled.

Neither the high school principal nor the director of public health has data at the person level for income, a potentially highly skewed variable. The bulk of the principal's and the director's data is not so skewed as to need special treatment. In such cases, statisticians breathe a sigh of relief, because statistical reports should be relatively simple rather than presenting puzzles of interpretation.

Along with skewness, another distributional characteristic deserves mention—*kurtosis*. Kurtosis refers to the *peakedness* of a normal curve. This issue is one that is best left to statisticians but is mentioned to familiarize readers with the term. As do large amounts of skewness, large amounts of kurtosis need special handling and can result in practically impenetrable interpretations in the results. Generally, skewness and kurtosis are ignored in social research. It is unclear to what extent this approach has resulted in the systematic misrepresentation of data or what impact such misrepresentation has had on the accuracy of world knowledge.

12. Averages—Central Tendencies

Mean, median, mode

What's an Average to do?

Ask the data

M any students first come to dislike statistics when introduced to central tendencies. This is no surprise. The topic begins by making something complicated that should be simple: an average. First, it gets a special name, *central tendency*, and three very different choices (*mean*, *median*, and *mode*, covered later in this chapter). Many students are then required to calculate several of each, as though struggling for a few hours will make them feel better about these concepts.

The choice of a measure of central tendency is determined largely by the structure of the data—mostly level of measurement and, to some degree, skewness. Here, researchers find themselves in one of three positions:

1. They know what they are doing and choose the correct average for their data.
2. They think they know what they are doing, so their choice is somewhat left to chance.
3. They know that they do not know enough to be sufficiently sure that they are making the right choice, so they get some help.

There are other combinations, such as those who know that they are less than optimally informed but go ahead and pick a measure. Most people doing statistics, however, care about the integrity and interpretability of their results, and do not want to mislead others while embarrassing themselves.

Strategic times for statistical help in research projects include, at least, at the conception of the project; during preparation of the data for statistical use; in special cases for which one really should have a statistician running the statistics; and in situations requiring precision in interpreting the results. Although both the high school principal and the director of public health have been known to cut corners when doing their research, they do so with the knowledge that they are more likely to have to justify or restate their results when they cut corners than when they involve more experienced statistical and research help.

12.A. Mean

The mean gets bullied
Pulled to the strongest side
Helpless in the tug-of-war

The *mean* is the arithmetic average, has intuitive appeal, and is the most common incarnation of the average or central tendency. Yet it needs a fairly symmetric (i.e., nonskewed) distribution to be a relevant approximation of the central tendency. Large numbers of outliers (data points far from the mean), or even a few data points extremely far from the mean, can greatly distort the mean. When the data are fairly well balanced, the mean is quite useful. It contains more information than the *median* or the *mode* (to be discussed in the next two sections) because it is affected by both the number and the size of all occurrences. Means also are quite handy for variables that take only two values, such as gender. As seen earlier, if the two values are coded 0 and 1, the mean is the percentage of the total that is coded as 1.

The high school principal will be using means for scholastic data. For example, if he wants to see whether grading differences exist across his various teachers of the same subject, he could compare the mean grades for different sections of the same courses for similarly academically inclined students. If large differences were found, he might want to talk to the teachers about resolving them.

The director of public health will use means extensively. For example, she will look for the average (i.e., mean) number of immunizations and various kinds of important medical services. Where differences are found in the means, she will look for associated conditions that might be addressed by policy.

Nonetheless, the mean can be a difficult concept. Let us say that a group is 60% male. What does it mean to say that the average person in the group is 60% male? When one foot is in scalding water and the other in ice water, how consoling is it that, on average, the temperature is fairly comfortable? Think about interpretations of the mean.

12.B. Median

Medians play the center

Insensitive

Unmoved

Medians are rarely used by choice. They are simply the point where half the values are larger and half are smaller. Needing only an ordinal level of measurement, they are a good choice for perception scales and for highly skewed data.

Medians are often paired with other information about the distribution—percentiles (hundredths) of the distribution. The 50th percentile is the point where 50% of the scores are below it and 50% are above it—the median. Often, the 5th, 25th, 50th, 75th, and 95th percentiles are shown as a way to characterize an ordinal distribution. Surprisingly, interval and ratio data also are characterized this way in many settings, even though other statistics exist to describe such data. This appeal is known to conference presenters as well as journal article authors.

Both the principal and the director will be using the median for data gained from perception scales. In addition, many of the principal's standardized test scores are best displayed as percentiles. The director also will use percentiles to explain the distributions of immunizations and other services for various groups in her state.

If a researcher had to live with only one form of central tendency, a good argument could be made for that choice being the median. In a symmetric (balanced) distribution, the mean and the median are the same value. To the extent that they differ in value, the median is more meaningful and the mean less so for many situations. Why? The reason is that the less symmetrical a distribution, the more the mean represents the impact of a few large, or many small, outliers. The median is, therefore, the more stable of the two forms of central tendency discussed so far. It does not require a symmetric distribution, so it functions well with either parametric or nonparametric data.

12.C. Mode

Strength in numbers
The mode boasts
But is seen little

The *mode* is the value with the most frequent occurrence. For data that take a large number of very specific values (e.g., length of each road in a state), the mode is not useful. For data that take a more limited number of values (e.g., make and model of registered vehicles), the mode can be informative and useful (e.g., America's 10 most popular cars). The mode can be calculated for the majority of distributions, yet it is the least informative of averages, or central tendencies. That relative lack of information content is due to the mode's failure to indicate its relationship to points anywhere else in the distribution.

Moreover, we have to guess at the mode's relative contribution to the entire distribution. For example, you would expect the mode to be 0 for the number of automobile accidents a typical driver might have in a 1-year period, or 12 for the number of eggs in a dozen that survive the trip home from the store.

Both the principal and the director will use the mode as little as possible and hope to avoid it altogether. It is the central tendency of last resort and the one that carries the least amount of information, and it is most contingent on artifacts of coding and methodology. Others might claim it to be simple to explain and easy to see on a graph.

The high school principal remembers an energetic discussion with a parents' group after he tried using the mode to describe the average grade at his school. Parents simply didn't understand it. The director of public health remembers the governor's speechwriter calling at midnight to get her to replace a section that referred to modes in a report the governor was to present the next day.

13. Two Types—Descriptive and Inferential

Knowing this

Projecting that

Reach can exceed grasp

The two basic branches of statistics are *descriptive* and *inferential*. Descriptive statistics discuss the data at hand. No projections are made to larger or other groups. Meaningful comparisons are not made, because doing so directly would not accommodate sampling error. Yet a fundamental goal of statistics is to be able to project findings to larger groups or to compare groups. When inferences are made about overall populations from the findings in samples, or a comparison of group characteristics is needed, we are using inferential statistics. As we do this, we accommodate sampling error and generate educated guesses by way of p values (described later in the book, specifically in Chapter 31).

Simply put, descriptive statistics describe data, and inferential statistics make inferences from the data to other situations or conditions. Most research uses both types. First, descriptive statistics are used to present the salient characteristics of the data so that readers have the background necessary to make informed decisions about the results. The population is described as well as is possible. Then a sample is described that a reasonable person probably would think represents the population well. When certain aspects of the data show larger differences than others, a report normally will contain text that recognizes and downplays the difference. Next, in the more technical statistical portion, inferential statistics are used to place the research in a larger and more important context than just the particular one at hand. This section of a report is critical for publication in many top journals and can influence one's chances at continued or independent funding.

Most people refer to central tendencies and grouped frequencies as descriptive statistics. Means, medians, modes, sums, ranges, percentiles, standard deviations (discussed later, specifically in Chapter 26), and sometimes skewness and kurtosis are all types of descriptive statistics. They all describe different aspects of the data.

Most inferential statistics involve statistical tests. Questions about, say, whether a rate of flu immunization for the state was above a targeted goal or was higher than that of a neighboring state are types of inferential questions. When you see a statement about something being statistically significant (or not), understand that inferential statistics likely were used. The reason is that statistical tests relate to a null hypothesis (covered later, specifically in Chapter 28), which is a statement about a population. The inference comes from relating the results from the sample to the hypothesis about the population.

The high school principal has questions about average grade differences in various sections of English and other courses taught at his school. He also questions how the athletes perform in their coursework compared with nonathletes. These comparisons are inferential because he generally will have samples rather than full information due to the length of time it takes to get some of the data into the electronic form he needs for access. He will use descriptive statistics in discussing the makeup of his school, such as in an annual report and in his presentations to parents and the school board. As we can see, it is through the inferential side of statistics that the guesses are made. Inferences contain an element of uncertainty both in everyday English and in statistics.

The director of public health has a more fluid population but still uses both descriptive and inferential statistics. She still believes that there is inherent sampling error in all of her data, but in several of her reports she describes her data as snapshots in time. Due to the large number of people's records available to her through electronic databases, she believes that her data are mostly representative of everyone coming into and leaving the state's system.

Almost all of the director's work in public health involves making inferences about the effects of public policies on people's lives. Even when she groups people by age ranges, she lists likely error ranges for her statistics. She has observed instances in which people remembered trivially sized differences and inferred that one group was doing better than another, and these same people then being surprised to see the relationships reversed in another report. By showing error ranges (i.e., confidence intervals, discussed in Chapter 35) from an inferential perspective, she protects the integrity of her department's work. Later on, when small changes in sample statistics could create small shifts that reverse the relationship between two statistically equivalent numbers, she can return to the original report and show how the changes are consistent with the amounts of likely error shown. Given the protection that can be derived later from showing likely error ranges (through, say, confidence intervals) in calculated statistics, one wonders why more statisticians do not use them everywhere that such an interval can be estimated. Confidence intervals from inferential statistics can be a joy for a statistician. Even when a later statistic falls outside the range given, the reply can simply be: "That range was for the 95% confidence interval. This event apparently is somewhat rarer."

By reporting inferential statistics properly, personal reputations can be saved and the integrity of statistics preserved. With inferential statistics, one cannot be 100% certain. In fact, language that appears to reflect certainty generally is a misrepresentation of the statistics. With a large sample, we can be quite certain, but not *absolutely* certain. This is a frustrating area for people who want clear-cut answers. Statistics are rarely clear-cut. Try not to fall into the trap of sounding more certain than your statistics can support. The way of statistics is a life of "probably" or "probably not," rather than of "yes" or "no."

14. Foundations—Assumptions

Magic . . . a surprise

What was thought is not

Ignorance has teeth

M uch of the math behind statistics is based on and supported by assumptions about the data, such as their level of measurement and how they are distributed. Some statistics have more assumptions than others. Some assumptions are more important under certain conditions than others. Understanding the role of assumptions means appreciating their limits.

One of these limits is the tolerance of a statistic to an assumption not being met. Statistical assumptions are technically violated, in one way or another, in the vast majority of statistics involving people. The important point to remember is that assumptions need to be met only to an appropriate extent, within the robustness of the statistic. Generally, the more complex a model is, the more tenuous the assumptions tend to become.

Remember to check the assumptions of the model, as well as your own. Be skeptical of statistics that are wrapped up as nice clean packages. With statistics, whenever you turn over a rock, you find dirt. Thankfully, a characteristic called *robustness* generally washes the model clean enough when overall care is used in the process. Violations of assumptions require decisions along judgmental continua. In some cases, talented people can disagree on those decisions.

Both the high school principal and the director have been "bitten" by some of their complex models when they have not fully understood the total package of assumptions. Neither is worried about simple comparisons, but when they get into model building, they will give their local statistician a call. Where the director of public health is a bit more concerned is in the assumption she makes about the representativeness of her data. Some of her measures may be based on small samples, possibly because of missing data for some people, and the missing data are poorly understood. Generalizing to the entire population can be tricky under these conditions. Statisticians can help untie these knots, but only within certain limits.

15. Leeway—Robustness

Data are rowdy

Messy, too

Tolerance prevails

S urprisingly, statistics generally are quite hardy and easily survive minor violations of their assumptions. This situation (referred to as *robustness*) is fortunate, because few situations conform entirely to assumptions. For example, real-world data do not conform to precise, normal curves, yet the normal curve supports much of statistics. Assumptions and the magnitude of their tolerable violations are the fodder for much joyous debate among statisticians. This issue becomes particularly rich in opinions when the discussion turns to aggregating assumptions (such as assumptions for the level of measurement combined with those of a statistical technique) for complex models. No one has this problem figured out or has a reasonably overarching approach to it. As a result, many statisticians have their own individual opinions of how closely the data need to fit all of the various assumptions behind a complex model so that the results do not become potentially too misleading to be usable.

The robustness of many statistics (to at least minor violations of their assumptions) is why both the principal and the director feel confident about running their own statistics for simple comparisons. From past experience, they know that their data are sufficiently close to what is needed to meet the various assumptions of their simple models. Currently, their biggest assumptions are that the overall structure of the data and the validity of their measures have not degraded since the statisticians last looked at them.

The degree to which researchers rely on the robustness of their models and of the statistics they use is quite remarkable. Because assumptions are never fully met with real-world data, researchers must constantly ignore them and rely on robustness to support their results. Therefore, researchers tend to grow somewhat complacent about examining assumptions that they know will not be met fully. Being aware of how assumptions and overstretched robustness can affect results is a good first step in developing a sense of when research might be faulty.

16. Consistency—Reliability

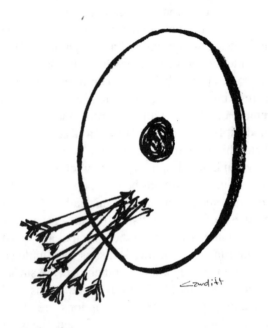

Again and again
Arrows tightly bunched
Where?

R *eliability* is repeatability, or consistency. It forms a continuum, despite people's fondness for talking about it in absolute terms. Having sufficient reliability is not the same as having sufficient *validity* (i.e., validity is hitting the bull's-eye, or "truth," and is covered as the next topic). Even when everyone might agree on something, everyone could still be wrong, as science so often demonstrates. Although validity cannot exist without reliability, reliability can exist on its own. As we will see later, establishing validity establishes reliability, whereas establishing reliability only establishes reliability itself.

Think of three different archers shooting arrows at three large targets with small bull's-eyes in their centers. The first archer's arrows land all over the target. Some arrows are 2 feet away from others. His shots are not reliable because they are not repeatable, not consistent, not all in roughly the same place on the target. Furthermore, because he missed the bull's-eye so often, his shots are not valid either. The second archer's arrows are all within an inch of each other. Some arrows even split others in finding their mark. This tight grouping of arrows, however, was 4 inches below and 3 inches to the left of the bull's-eye. His shots were quite reliable, forming the tight bunch as they did, but they were not valid, because they missed the bull's-eye. The third archer's arrows are just as tightly bunched as the second's but are almost exactly in the middle of the bull's-eye. Her shots were both reliable (consistent) and valid (hit their intended mark). No means exists for an archer's arrows to be valid (hit the bull's-eye) but not reliable (not tightly bunched).

One method of estimating reliability is to have two or more measures of the same construct and compare the results. These multiple measures can be multiple people "scoring" one thing, or they can be multiple scores for the same people. Either way, repeatability or consistency is being assessed. No estimation is made for the degree of correctness of the decisions, for that would be an estimation of validity.

Suppose two people were to score the same test, comprising open-ended questions about American history. The graders can score a question only as correct or incorrect. When we look at a cross-tabulation of their scores for each item on the test, four conditions exist. They could both agree that an item was answered correctly, they could both agree that an item was answered incorrectly, the first could score an item as correct and the second could score it as incorrect, and vice versa. The graders would agree with each other half the time if each were randomly scoring the items. Knowing that they agreed half the time is the same as if they agreed randomly or as if each had flipped a coin to determine truth. Statisticians responded to this situation (give them a challenge, some time . . .) with a statistic that controls for estimated random agreement, the *kappa* statistic. This statistic is a measure of reliability after removing the random-agreement component.

The high school principal knows that final grades are based on many quizzes and tests, so they probably are fairly reliable. He is not as sure about his data for socioeconomic status. Those data are from a single measure on a parental survey. The principal knows that he needs reliable measures to have credible results and will avoid measures based on single measurements. Because he knows that errors of estimation tend to cancel, he prefers aggregate measures that have had a chance to let several measurement errors mutually reduce themselves. In most cases, multiple measurements of the same characteristic or construct are more reliable than just one.

The director of public health has some doubts about the reliability of her data for medical screening of children for hearing, vision, and dental problems, because the documentation for some groups and from some clinics is quite poor. This issue shows where the line starts to blur for some people between reliability and validity. The director first doubts the reliability of the records because some medical record abstractors would have shown more screening being done than what is currently shown by others. She also doubts that the records accurately represent the degree to which some children are screened because she believes that their records are missing important screening being done in the schools—that they have poor validity. Reliability and validity are related (as will be shown further, in the next chapter).

For most of her data, the director of public health believes that independent sources looking at the same information would come to roughly the same results. In other words, she believes that the rest of her data are sufficiently reliable for her purposes without added cautions. She might still use the medical screening data, but she would do so as more of an example of why better medical services documentation is needed.

Reliability is a judgment call that is based partially on personal experience and partially on the historical use of similar data. Guidelines exist but are somewhat of a contrived method for handling the issue. Researchers need to resolve for themselves the required level of reliability that makes sense for their situation. Finally, researchers need to describe the levels of reliability in their work so that readers can judge for themselves whether the results are likely to be sufficiently repeatable.

17. Truth—Validity

From the mighty source

Final authority flows

To judge the flights of arrows

From the flex of bows

S everal types of *validity* try to answer this question: To what extent are we actually measuring what we think we are measuring? The various methods that statisticians have developed in attempting to answer this question are quite remarkable and varied. When questioned, people seem to think that they agree on what they mean by validity. That agreement breaks down, however, when you ask them to describe it in some detail.

This lack of agreement is what has led to many methods for assessing validity. Each method assesses a somewhat different aspect. For example, *criterion validity* assesses the extent to which a given model reproduces a given set of results. *Predictive validity* assesses the extent to which a given set of measures can accurately forecast a set of results. Both analyses could be done with the same database structure and would produce the same numbers as results. The interpretation of those results, though, would be different. This difference is due to the substantive context that produced the data in the first place.

Trying to develop consensus on the structure of validity might be a good first step. Others have outlined it, some in considerable detail. The idea is that once we agree on its structure, we can then try to design better ways to assess it. Some statisticians have suggested that all forms of validity are special cases of *construct validity* (the degree to which we are actually measuring and assessing what we intend). I suggest that this must be true, insofar as how the terms are defined. Construct validity can comprise all aspects of an issue, analogous to the flight of the arrow starting from the ride in the quiver and ending at its removal from the target. With this definition in place, any given aspect of the arrow's journey that forms its own type of validity can be viewed as a special case of construct validity.

Here are a few of those special cases. *Face validity* asks whether what we are measuring rings true on its surface to an expert in the field. *Concurrent validity* asks how well what we are measuring agrees with a standard for the same trait. Criterion and predictive validity were introduced earlier. These concepts result in mathematical equations but are not independent traits of the data. Rather, validity is the intersection of *intent* (i.e., what we are trying to measure) and *process* (what we actually measure). Various aspects of this intersection form the basis for various types of validity.

With that, mathematical relationships between reliability and validity support the notion that sufficient validity presupposes sufficient reliability. Philosophically, the results are the same. One cannot keep hitting a bull's-eye without being consistent. The reason why researchers cannot simply assess validity is that frequently there is no way to independently know the truth of the matter, the so-called correct measurement. Some fields refer to measures of reliability that involve a "gold standard" as a measure of validity. Under the assumption that the gold standard is in fact always correct, reliability will transform itself into validity. That assumption is sometimes impossible to guarantee. To the extent that it is in error, measures of gold standard validity stay at the level of reliability. Most readers have little means to determine the extent to which a gold standard actually is one.

The high school principal does not have to worry much about the validity of most of his data. Grades are designed to reflect academic achievement. Sports scores and club memberships are pretty clear. A few things, though, such as income level (as previously discussed), might not be sufficiently valid. Part of the reason is the reliability issue previously stated. The validity issue comes from the categories used to group income. Those at the top of the category and with little debt might have very different life experiences using their income compared with those at the bottom of the same category who carry much more debt. Then again, self-reported income can be somewhat suspect.

The director of public health has been looking hard for years at the validity of her measures. For just about every report showing differences in needed services received or the extent to which good clinical outcomes are met (by clinics, hospitals, etc.), those facilities or communities showing poorer performance claim the data are not valid for one reason or another. She has learned where her data are strong and where they are weak. Importantly, she learned a long time ago not to hide or ignore those aspects of her data. Her data are not a personal reflection on her unless she makes them that way. By standing independently and reporting the strengths and weaknesses of her data and results as impartially as she can, she remains a neutral party in later discussions that might arise.

Here is one final thought, almost a meditation, on validity. Validity rests on the nature and on the strength of the evidence in a given situation. Think hard about evidence: What is it? What is needed? Thinking about types and levels of evidence results in understanding more about statistics in general. To determine the validity of a measure, you must first decide on the nature and on the strength of the evidence needed to make the judgment. You will find that what you need depends on the situation and cannot be easily codified into a set of rigid rules.

18. Unpredictability—Randomness

Patternless, capricious
Noise abounds
Carefully depend on it

R *andomness* is the quality of being unsystematic or haphazard. Random samples tend to represent their parent groups fairly well, though, because everyone had an equal chance of being included. Randomness implies being unbiased. So, in the long run, all segments of the parent distribution will be appropriately represented. Random surveys tend to be pretty informative because their data are representative and, thereby, are informative, as long as a representative portion of the population returns them. To the extent that samples are not random, their data are not representative. When a representative sample does not seem likely, it is time to expend resources on a statistician and still not expect miracles. Certain types of sampling can overcome issues of representativeness known in advance of conducting the study, but issues arising during a study are harder to accommodate.

Many researchers handle this issue by being somewhat humble in generalizing their findings. They characterize their samples and methods as clearly as they can. They let their readers decide whether the findings can be generalized to the readers' populations of interest. More than a convenient trick, the approach is a genuine disposition for how knowledge is best generated.

Both the high school principal and the director of public health have the majority of records for their populations. Furthermore, when samples are needed, making them representative should not present any particular barriers. Besides, *p* values help to sort out the remaining uncertainty.

In general, random error is presumed to cancel itself out in the long run. Other assumptions on randomness vary from statistic to statistic. One way to think of it is that the random error inherent to statistics is part of the uncertainty that allows statistics to exist. Nonetheless, statistics and larger samples cannot undo the problems created by poor methods of sampling (covered next, in Chapter 19).

19. Representativeness—Samples

Sampling's goal?

Close enough

Resources rule

S ampling is a statistical response to limited resources, an efficient way of estimating values (i.e., making educated guesses) for large groups. Several families of sampling techniques exist; each is designed differently to handle varying conditions for optimizing resource use. A few general rules apply across the various sampling designs:

- More representative samples yield better results.
- All else being equal, larger samples yield better results.
- Larger samples cannot make up for a poor sampling plan or for poorly executing a good plan.

For example, there is the story of the U.S. Census conducted in one major city that had a "corner problem" with undercounts as a result. The door-to-door interviewers were paid by the number of streets covered. Buildings that were on the corner of two streets more often were considered someone else's responsibility than buildings in the middle of the block, for obvious reasons. The result was an under-representation of people who lived in a corner building. Now, was that lack of representativeness big enough to hurt any analyses? Maybe. Areas with proportionately more corners per square mile might see a disproportionate underrepresentation than would other areas. State funding for road and utility construction, based on population counts, could be affected importantly by such an undercount. Representativeness both of the overarching population and of its subcategories matters in samples as well as in sampling.

Now to the main question in sampling: How many subjects do I need to . . .? The answer is rarely a single number (although "probably more than you can afford" likely is correct but unwelcomed). The reason is that big differences are easier to find than small ones. People know this and want to find big differences but, when pushed, ask to find smaller ones.

How big (or small) a difference do you want to find? For example, how large a sample might be needed to see whether a professional football player could break a glass bottle with a baseball bat? One. How large a sample might be needed to see the extent to which thinking "light" thoughts could result in weight loss? A lot more than one. Big things are easier to find; small things are harder. The rest of sampling is a continuation of that simple realization. The most difficult part of sampling for most people is determining how large a difference (or change) is needed to be considered as substantively important. Once that is determined, the next issue is whether resources are sufficient for obtaining the necessary sample size.

Determining the specifics behind a choice in sampling methods and the required sample size are topics that are well left to statisticians, who debate the topic easily and often. Much of the process is driven by equations (several are available from which to choose, further livening up the debate), but the process also is driven by judgmental accommodations of the unique conditions existing at the time. Moving quickly from the tidy examples in sampling textbooks to the world of applied research, one finds that many of the conditions and assumptions in the texts are absent or contradictory in the field.

The result for most people is that just about any sampling methodology beyond a random sample probably should involve a statistician. The methodology becomes complicated quickly, and the statistics have to disentangle the nonrandomness before being substantively interpreted. Nonetheless, many sampling issues do not matter that much when using electronic data that are captured (theoretically) for everyone in a population. For the high school principal, even his surveys could be sent to everyone who would be or was affected by whatever he would like to study. The director of public health knows that nesting (covered in Chapter 42) is taking place in her data, and she brings in her statistician early in the research process. An example of nesting is when neighborhoods are divided along ethnic lines. Additionally, the completeness of the medical record varies by county, so ensuring representativeness for the state is quite complex. She finds that understanding sampling theory is much like holding sand in her hands. When she just picks it up, she has it. After carrying it around for a while, she finds that much of it has slipped through her fingers and is no longer with her.

Many people find statistical concepts seem to vanish fairly quickly, as did the sand. Statistical concepts are different from day-to-day life in that they are probabilistic. People like to believe that things happen or not, rather than that they probably happened or probably did not. In fact, in one branch of statistics, item response theory, a full point is not awarded when a test item is correctly answered, nor are zero points given for an incorrect answer. Probabilities are calculated and awarded for achieving the correct answer for each item and are summed to arrive at a total test score through yet another algorithm. Gone are the days of adding up the number correct and dividing by the total number of items. Now, you get some credit for items answered incorrectly; you do not get full credit for those answered correctly; and on some tests, you get partial credit for items that were not even on the test you took. Sampling and samples have taken us a long way down the statistical path—some people might suggest too far down it. Others just stand back and watch in amazement.

The goal of sampling is representativeness. Larger samples yield more precise results than do smaller samples. These two simple truths arise again and again.

20. Mistakes—Error

Ambiguity and approximation

Cornerstones for existence

Truth pokes through the middle

Error takes three basic forms in statistics: sampling error, measurement error, and estimation error. *Sampling error* is the degree to which the parent population is not faithfully represented by the sample. *Measurement error* is the amount of the observed score that is considered random noise and not part of the measurement. *Estimation error* is the window surrounding statistical estimates that indicates the likely range for those values, given the data at hand.

Sampling error is reduced with scientific sampling techniques and larger samples. Measurement error is reduced with strategically designed, psychometrically assessed, repeated measures designs. Estimation error is reduced with larger samples and better models.

Errors of measurement tend to cancel with multiple measurements. Although measurement error degrades the precision of individual values, estimates made from many measurements tend to be pretty dependable. The amount of error remaining after aggregating several measures depends on many issues. In general, the more people or measures in the study, the less error will influence the results. Additionally, in relatively small populations, the larger the percentage that is studied, the greater the accuracy in the answer (but this tendency breaks down in large populations, such as in cities).

The high school principal has a high percentage of his population for all the measures in his data and has multiple measures for most things he wants to study. He is not too concerned about measurement error. The director of public health also has a high percentage of her population in her electronic data, but her surveys are only of a relatively small percentage of those eligible for services. Fortunately, after a certain point in sampling, little is gained from sampling more, independent of the population percentage it might represent. An important fact about sampling is that its precision follows a curve of diminishing returns.

21. Real or Not—Outliers

A snake!

Coiled rope

Bitten anyway

O *utliers* are data that are outside the typical range of values for a given situation. They can be real, or they can be one of many types of data errors. The idea behind outlier removal is to figure out which of the data are likely to be genuine and which are likely to be spurious. Real data that do not correspond to what typically is expected can contain valuable information on how they got there. This type of information helps us to understand the true drivers of complex systems.

Deciding which data are real and which are likely to be mistakes is frequently reduced to a series of educated guesses. Importantly, these seemingly minor decisions about outliers can produce large changes in the results of the analyses. Equally important, these decisions are rarely documented in reports or journals, leaving readers to trust with little evidence for support.

Should you discard outliers and use a cleaner but often less representative and less informative data set? Or should you leave the outliers in the data set and risk mistaken data influencing the results, sometimes greatly? You are faced with two undesirable alternatives, but that is the nature of data (*Bitten anyway*).

Experience shows that handling of outliers is often pragmatic. The high school principal and director of public health face similar issues and make similar decisions. They both compromise by trying to find a middle ground between accepting too many potentially biasing mistakes and discarding potentially important information. They then trust to luck. They hope that their methods of compromise result in their being only nipped, and not actually bitten, by their decisions. Although mechanical rules of thumb exist, they quickly break down when retaining the true information content of the data is the driving force behind outlier (used here as *mistake*) removal.

22. Impediments—Confounds

A suspect was captured

His conviction assured

While safely at home

The real thief sipped tea

*H*ow often have we read findings from one research project, only to find exactly the opposite in print only a few weeks or months later? The reason often is confounds. *Confounds* are variables that might really be responsible for variation in the data that interests us but whose effect has not been accommodated. That is to say, a confound is a variable that is related to both the dependent variable and the independent variable of interest but whose influence has not been ruled out, or accommodated.

Three ways exist to accommodate potential confounds. The first and best is through the research design, where a good design all but precludes plausible confounds. The second is through statistics, where the impact of these variables is statistically controlled. The third and weakest approach is by appealing to logic and argument: the "it stands to reason that . . ." tactic. Few journals accept the third technique. Most reputable information outlets (journals, books, national media, etc.) prefer the first method but also will accept strong examples of the second. The reason that they want truth, as much as researchers do, is that it eventually catches up with them anyway, sometimes uncomfortably.

Confounds are the missing ingredients that are waiting to catch up with each and every researcher. Confounds lead to "why didn't I think of that?" moments. Moments when confounds eventually are discovered are when researchers find out that embarrassment can be an unfortunate part of a productive professional life. No matter how hard researchers try, they cannot think of everything, every time. The situation is much like losing something rarely used. The owners of the property can go months, years, and sometimes forever without noticing the item's absence. Thinking of items that should be in our models but are not is more difficult than seeing items in models that do not belong there. Models are based on theory and experience. History has shown both to be vulnerable to additional information.

Good research designs are the first and the best line of defense against confounds. Good research designs tend to isolate whatever is being studied. This isolation conveys a sense of confidence that the intervention or experiment was responsible for the ascribed changes in the dependent variable. In the real world, however, isolation is hard to accomplish, and full knowledge of the influences on a given situation is difficult to achieve.

Most complex models are designed to statistically accommodate recognized confounds. In fact, many of the more exotic techniques were developed for the purpose of controlling for confounds under various conditions. These techniques remove the impact of the hypothesized confounds on the results. The twin issues for thinking about confounds are first, what those confounds might be, and second, how to find suitable measures for them. As hard as those tasks might sound, they are likely even harder to accomplish in practice.

Both our researchers have previously retracted some of their findings due to later-discovered confounds. Neither would like to repeat the experience. The high school principal once found that athletes were rarely tardy for school. Then he found out that teachers were reluctant to mark them tardy because they would be ineligible to play that week. He found out because other students were heard complaining about the unequal treatment from homeroom teachers. Not only was the principal embarrassed at having the wrong findings, but he also was angry that teachers would differentially enforce school policy. The discovery of the confound revealed a problem at school.

One reality of research is that you can neither think of, nor can you control for, everything that might be responsible for some of the findings that were ascribed to something else. To advance knowledge, we must be willing to be wrong, as hard as we try to avoid it. It is often from showing where others were wrong that knowledge achieves its largest gains. Sometimes we are the ones who are wrong—another good reason to be humble in the presentation of results.

Imagine, for a moment, being a newspaper reporter and turning in the following story. The headline is "Math Instruction Helps Verbal Achievement Scores." The reporter found that high school seniors who had taken 4 years of high school math got higher scores on standardized math exams than did seniors who had elected to take only 2 years of high school math. Although the higher math scores did not surprise the reporter, he noticed that the students had higher English scores, too. He concluded that the extra benefit to English achievement from taking more high school mathematics should be an important factor in trying to influence more high school students into taking more mathematics. The embarrassed reporter later realized the confound. The students who elect to take 4 years of high school math are the smarter, high-achieving students anyway. They would have scored better on both tests even if they had not had the extra years of math.

The revelation of a confound can be a forehead-slapping moment. They tend to be "Why didn't I think of that?" types of moments. In a sense, the ways in which we handle confounds, uncovered before or after our research results are made public, says quite a bit about our character, our inner selves. Handle confounds with humility and grace; but more important, handle confounds with a sense of inevitable humor, because you cannot be lucky every time.

23. Nuisances—Covariates

What was the cause?

How do you know?

Reconsider!

C ovariates are traits or issues that might interfere with arriving at fairly correct results. They are the statistical incarnation of what would be confounds if they were not accommodated. Covariates help to answer to the question "How do you know it wasn't such-and-such that caused that-and-which?" Those questions often are excellent, and good research is required to be able to answer them appropriately. Covariates, then, are those characteristics that need to be adjusted out or controlled when we use statistics to answer our questions in all but the most contrived environments. They are used in such methods as analysis of covariance, multiple regression, discriminant analysis, and canonical covariance analysis (all covered later).

Mostly, covariates spring from not having the research luxury of truly random assignment to groups. In real-life research, it is rare to have truly random assignment, where the groups are equivalent for all intents and purposes. Covariates are used to try to overcome that handicap by making adjustments for the original differences between groups. No matter how many covariates are used, the results still are not as strong as if truly random assignments were used and everything else but the factor being studied remained equal for the groups to be compared. Although covariates are imperfect adjustments, in a very real way, they make the vast majority of research possible and sufficiently credible.

Covariates are strong medicine. They need to be used sparingly in all but very large data sets. Data themselves are not perfectly random. Covariates tend to capitalize on the nonrandomness of the data and overfit the model. If an overfit model were applied to a new sample, the results would be sufficiently different to make us question whether both samples came from the same parent group. Depending on the importance of the results (e.g., how much and how many people will be affected), many researchers want to see from at least 10 subjects per covariate (e.g., for lower stakes, student types of projects) to at least a few hundred (e.g., for interventions in public policy that could affect public health).

Recall the three ways of overcoming confounds: good research designs, statistical controls, and appeals to logic. Covariates are the fodder of statistical controls, insofar as they are the keys to effecting statistical controls. In supporting the validity of research results, the seemingly limitless number of questions of the nature "Did you think of . . .?" "How did you control for . . .?" and "Couldn't it have been . . .?" are by and large addressed through the thoughtful use of covariates, plus a good research design.

The high school principal probably will need to think hard about using covariates. Some of his comparisons will have to adjust for academic ability. For example, if comparisons of average (i.e., mean) grades for scholastic club members versus nonmembers are made to see the effect of clubs on grades, then the comparisons would have to adjust for prior differences in academic achievement. Prior grades, prior course difficulty (i.e., college prep or not), and standardized test scores could be used as covariates to make these adjustments. The principal's biggest problem with covariates is that he has too many of them for the number of students he has at any given time. Even a large classroom has 30 students. Some of the honors classes have as few as 17 students. He must pick and choose to avoid overfitting his models.

He looks for natural experiments: situations that are automatically conducive to research. These situations exist where, say, about 50 students were divided, based on last names, into two sections of the same course. Every other last name went into Section A, and the others into Section B. The principal has just such a case in the upcoming American History course. The materials, resources, time of day, and placement in the building of the two sections (they are next to each other) are all the same. Even the tests are the same, multiple choice with no room for subjective grading. The only difference that he can see is that one of the sections will be taught by an inexperienced teacher, whereas the other section will be was taught by one with more than 20 years of experience, who has won Teacher of the Year seven times.

With this wonderful natural experiment in hand, the principal could test for a difference in average (mean) grades between the two sections. No covariates would be needed, because all of the starting conditions were as close to equal as should be needed. In effect, the two course sections were academically equivalent at the beginning of the course. Given this natural experiment, the principal could use a *t* test (covered in Chapter 39) to see whether the difference in teachers' background and experience was related to differences in academic achievement in the American History course.

The director of public health will use some covariates, even though most of her measures are for clinical services that everyone in her population should receive. She has found through experience that inappropriately used covariates can remove the effect of a problem that she is looking to find. The reason is that when one is looking to find the effect of one variable but controls for one that is moderately or highly related, the portion of the variable of interest that is related to the one being controlled is lost, as its effect has been removed along with the rest of the information in the variable. With that in mind, she will use some covariates but will be very careful about how.

24. Background—Independent Variables

Picture me now

A portfolio of views

Like never again

I ndependent variables are the groups and background variables that enter into questions about differences. For example, I might want to know whether Democrats and Republicans had different average heights. If so, their group affiliation would be their political party. We want to see whether group affiliation is related to differences in average height, our targeted issue. The variables that are the targets of our questions are the *dependent variables* (covered in Chapter 25). Independent variables, therefore, form two broad classes: variables of interest (often grouping variables that lead to answers to our questions) and covariates. In most cases, they are handled identically in statistical models and tests. The difference is whether we want to interpret the size of their impact or leave them as controls to unequal grouping conditions.

The high school principal will be using class, course, club, team, and demographic groupings. He will make sure that the levels of measurement are sufficient for the techniques that he will use. He will try to avoid *proxy measures* (i.e., indirect measures of a trait of interest for which more specific variables are not available) when he can. Just as covariates are imperfect controls, proxy measures tend to adjust more for their difference from what was really intended than they do for their original intent.

The director of public health will use several proxy measures as independent variables, although she would prefer otherwise. Many of her data are at an aggregate level (averaged over geographic or administrative areas). For example, she has data for income at the voting bloc level, although she would prefer it at the individual level. Nonetheless, by staying fairly humble in her results, she avoids much of the embarrassment that could result from treating this class of independent variables as though it represented more direct measurements.

25. Targets—Dependent Variables

Center of attention

Questions asked about me

I spark controversy and progress

D *ependent variables* are the central focus of our questions. They are the things that matter for whatever issue is being addressed, the targets of models. As such, dependent variables tend to be the measures that really interest people. Specifically, people want to understand which issues affect variation in the values of the measures for these variables, so that they can try to influence them. This understanding forms the very foundation for public policy and its research.

The level of measurement for a dependent variable can profoundly influence the choice of statistical technique in the more complex models. Think hard about how your dependent variables are measured and structured. A conversation with a statistician in the early stages of a research project can reveal opportunities to refine questions of interest to better match the specifics of the data, while still addressing the issue in question.

Here is where the two researchers greatly differ in their interests. The high school principal is concerned mainly with differences in the quality of education that his students receive in different programs and from different teachers. High school grades therefore will almost always be his dependent variable, especially for some of the more complex models, because differences in grades could be due to a large number of things.

The director of public health wants to find out where and for which public health issues her funds are doing the most good. Her dependent variables are a wide variety of public health measures, such as rates of vaccination for children and flu shots for elderly people. Some of her models will use dichotomous or nominal level data for her dependent variable. These types of data require special care in their use and in the interpretation of the results. She will need a statistician's help more than will the principal.

26. Inequality—Standard Deviations and Variance

Differences sing out

Trumpeting unique notes

Who's minding the score?

ifferences are important. They contain information. Knowing why something or someone is different is generally more important than knowing how they are different, although knowing how they are different forms the basis for the art of choosing the right statistical technique. The *standard deviation* measures the extent of differences in a metric (unit of measurement) that is comparable across measures. It does this by representing placement under the normal curve relative to the center of its bell shape. For IQ, for example, a standard deviation is about 15 points (the mean is about 100). Scores within one standard deviation from the mean, on either side, will include about 68% of everyone. Scores within two standard deviations cover about 95% of everyone under the curve. Scores within three standard deviations include just about the entire population. Beyond three standard deviations in either direction on a normal curve, the air gets pretty thin; events are rare. With large populations, such as for countries, even the tails of the bell-shaped curve (past three standard deviations) can represent large numbers of people. Nonetheless, proportionately, they are still a very small percentage of everyone under consideration, at least under the normal curve.

The size of a standard deviation, relative to its mean, represents the degree of inequality in a characteristic or trait. Relatively wide standard deviations suggest greater inequality for that measure than do relatively small ones. Standard deviations therefore present both information content and a measure of the degree of inequality in a measure. If a standard deviation is zero, that measure has no information content in statistics. For example, if our entire sample is composed of 6-foot-tall people, testing the impact of height on something else cannot be done because there would be no data (and no information content) for other heights; therefore, no frame of reference would exist from which to draw conclusions about the impact of height. The computer would give us an all but unintelligible error code, but the result would be the same. No variability means no information content, which means no frame of reference for drawing conclusions. Information content forms an important frame of reference for statistics.

I do not think it worth spending much time on why the standard deviation multiplied by itself (i.e., "squared") is another useful measure of differences, called the *variance* (actually, the square root of the variance is the standard deviation, but functionally we find ourselves in the same position). Although variance is useful to statisticians, it does not translate well into English. For example, variance in income would be measured in squared dollars. Variance in the sizes of groups is measured in squared people. What would a squared ping-pong ball look like? Although useful to statisticians for a variety or reasons, for most of us, thinking of standard deviations and variance as types of "information content" or "measures of inequality" works pretty well.

As measures of inequality, standard deviations and variances can highlight some of the best and worst in society. Although without intrinsic societal worth in their mathematics, their application generally is deliberate, which means it is subject to researchers' personal values. Once more, researchers need to be careful in how they apply their knowledge of the systems and data they have. Showing one group to score lower on a sensitive measure, such as IQ, can spark a debate that is hard to quiet. Various measures show various relationships, whether they fit our sense of civility or not. We all need to make personal judgments for where we will and will not be involved. Statistics, like science, is not value-free in its application, despite the protests of some who might like to see it otherwise. Questions about gender, race/ethnicity, social status, and many other features of daily life can lead to hurtful results when handled without due concern for topic and context. When we make a bomb, we know it might be used. When we present a few statistics that could be harmful to a particular group, we need to do so knowingly and reflectively and, maybe, quite rarely, if at all. The way of statistics, though, should be to engage in that debate while erring on the side of humanity.

The high school principal and the director of public health are somewhat alike in their understanding and use of standard deviations and variance. They both view them as measures of information content and, perhaps more important, of inequality. Inequality is not a value judgment in this statistical sense. It is simply a statement of what exists and can lead to interesting questions about why.

This point is well worth remembering and sometimes worth debating. Good research is unbiased in terms of personal values, in approach, coding, interpretation, and all the small decisions that enter into answering questions. Yet none of us is value-free. We are all products of our past, of our pervasive conditioning that permeates all aspects of our conscious lives. To assert that anyone can fully escape this conditioning is to make a classic mistake in research—arrogance. We cannot be fully unbiased. So, while we try to be as unbiased as possible, we must accept the possibility of just plain being wrong or being unconsciously biased in our approaches. We need to accept these things because our prejudices can influence our results, whether we like it or not. Being constantly aware of these issues helps us to keep on the path of statistical knowledge.

27. Prove—No, Falsify

Popular hero

Cheering support

Glass jaw

S tatistics do not *prove* anything. Statisticians gave up on that idea right from the start, or close to it. Why? Simply put, statistics are based on probability (from data from samples). Being based on probability and probabilities always being above 0% and below 100% (even if by a minuscule amount), statistics offer no absolute guarantees. Billions of supporting examples for absolute truth are outweighed by a single exception. Finding an additional example that *proves* something is really just finding another supporting case, not proof. So, in statistics, we can only try to disprove or falsify. In a very real sense, we try to find, in some way, the exceptions. When involved with statistics, absolutes tend to have glass jaws (notice: even that statement was qualified).

Having been sufficiently weathered by experience, neither the high school principal nor the director of public health uses the word *prove* in his or her reports or presentations. They talk about either disproving the null hypothesis (covered in the next chapter) or failing to disprove it. They will frame their questions such that they try statistically to disprove them—that is, find evidence of exception. They also know that absence of evidence is not evidence of absence. Just because we might not find something (absence of evidence) does not mean that it is not there (evidence of absence).

The issue of not proving things with statistics takes us into the realm of causality. Causality is a strange and awkward domain for statisticians. It makes them uncomfortable and activates a heightened sense of caution, as causality is quite difficult to establish with statistics alone. All manner of contrived conditions have been, and still are being, created to try to move statistics from what it does quite well all by itself (i.e., addresses questions about associations) into a more deterministic domain.

28. No Difference—The Null Hypothesis

"No difference," I declare

The starter's pistol sounds

Can a difference be found?

In time?

T he *null hypothesis* puts the question of interest to the statistical test. It is normally a no-difference, declarative statement, such as "The average math scores from two different teaching methods are the same" (i.e., are not different). The alternative hypothesis usually is the opposite: The two teaching methods result in different average test scores. The beauty of this approach is that in the end, only one of two statements is left standing: (1) The sets of measures are not statistically different and are, therefore, the same; or (2) the sets of measures are likely to have come from samples of different populations because they are statistically different. The first statement supports the null hypothesis, because the probability that the scores come from the same population is too large to reject the null hypothesis. The second statement supports the alternative hypothesis, because the probability is too small to suggest that the sets of scores come from the same parent population. Statistically, failure to show a difference means that the null hypothesis still is tenable. The research question at hand might complicate these decision rules a bit (e.g., with three or more groups), but the basic process is the same. A no-difference hypothesis is presented along with some alternative to it.

In *two-tailed tests*, the hypotheses suggest that the results could get better or worse, go up or down, or become higher or lower, in some way. A new way to teach mathematics could result in either higher or lower scores. For this situation, the alternative hypothesis would be that the new way of teaching mathematics results in different average scores. Remember, the null hypothesis stated that the average scores from the two mathematics teaching methods would be the same. Most statistical tests seen in journals are two-tailed tests. They are more conservative and generally are better representations of reality than are one-tailed tests. Two-tailed tests also are the default of most current statistical software. You need a pretty good reason to use a one-tailed test. Here is why.

For *one-tailed tests*, the alternative hypothesis is one-sided—for example, the new way of teaching mathematics results in higher average scores. The null hypothesis would remain unchanged. Notice that no mention is made of the possibility that the new method of teaching mathematics could result in lower average scores. Be wary of one-tailed tests. Although they sometimes are appropriate, one-tailed tests can also be a method by which results that are close to statistical significance can be made to look statistically significant using the same data and an appeal to logic. We already have covered the relative strength of an appeal to logic in the chapter on confounds (Chapter 22). The application of logic to statistics is no stronger here than it was for controlling for plausible confounds.

The high school principal is experienced with null hypotheses but does not use them for reports. His work is not of the kind that would be submitted to journals. He needs numbers for overhead projections presented to parents' groups and meetings of the board of education. When he uses a statistical test, he sees little reason to follow the tradition of specifying a formal null hypothesis.

Both the high school principal and the director of public health know that they have either supporting or disconfirming evidence, no more than that. This forever-tentative mindset is a toll that is paid repeatedly on the path of statistical knowledge. A tiny piece in the back of the mind keeps us from being absolutely certain of things. We try to remain tentative for two reasons: the nature of samples and the nature of epistemology itself.

The nature of samples includes the characteristic that they cannot be complete; otherwise, they would be populations, by definition. This means that some amount of unknown information about the population interferes with the notion of proof from a sample; that exception might be lurking out there somewhere. Besides, proof is just too certain for samples, and statisticians by and large enjoy the luxury of not being responsible under most circumstances for having the future turn out somewhat differently from what was predicted.

Epistemologically, what a given person knows depends on how it came to be known, something that varies from person to person. What we know of topics or events differs from our varying perspectives as well. If we saw an arrest, our concerns would be different if we were another police officer compared with if we were a visiting dignitary. Statistics are not different in this regard, nor are the judgments made by people for the conditioning of data (i.e., the intentional and strategic structuring and coding of a database to facilitate the analytic portion of research). The conditions by which we learn things and the specifics about them subject our knowledge and memories to constraints that have little to do with the data. We need to guard against these unique portions of our understanding somehow filtering into how our data are analyzed.

Results, therefore, can be quite dependent on the methods used in finding them, both in the conditioning of the data and in the statistics used to address the questions. This unintended consequence of people creating, directing, modifying, and independently interpreting subtly ambiguous aspects of research protocols too often is reflected in research results that fail to be replicated under conditions whereby substantially equivalent results would likely be found. This situation is most commonly found in statistical modeling (covered in the next chapter).

29. Reductionism—Models

Too much to carry

What is needed?

Ignore the rest

M *odels* are useful distillations of reality. Although wrong by definition, they are the wind that blows away the fog and cuts through the untamed masses of data to let us see answers to our questions. Models try to focus on the salient qualities important to answering our questions. Models reduce reality to an amount of information that can be handled. While researchers whittle through masses of data, they make many decisions that could alter the results of a study—so many decisions that they can rarely all be reported, even if documented at the time. Many of these decisions are later thought to be plausible confounds and are resolved in subsequent work, resulting in different findings. Nonetheless, good models can yield important insights into complex situations.

The high school principal will use many simple models, such as correlations or analyses of differences in average grades (i.e., differences in their means). He will do few complex analyses to account for the ways that students are grouped within course sections, within courses, or within any of such "nested" conditions. He knows that he needs to call his statistician when the situation being addressed is more complicated than a comparison of two or three equivalent groups. The reason is that the statistics and conditioning of the data can quickly become very complicated.

The director of public health finds that the state's beneficiaries are grouped mainly by neighborhood and access conditions. She realizes that many of the services in which she has an interest are intended for almost everyone in the population. These services (e.g., vaccination against pneumonia) are subject to variable delivery to the public according to the variable persuasiveness of providers. She needs to get providers to be more persuasive in certain geographical areas that are relatively easy to locate.

30. Risk—Probability

Not yes, not no

Between

Where?

How can we calculate probabilities that are closer to reality than random guesses would be? The history of gambling plays a big role in answering this question. Where there is money, there is motivation. Plenty of motivation could be found with coin-flipping, dice, and card games. People were making wagers and, hence, started figuring out the odds. They also asked friends for help with figuring out the odds. Add several centuries and mathematically inclined people with time on their hands, and you get modern statistical probabilities. Fortunately for most of us, understanding the meaning of that probability is pretty easy and does not change much among choices of statistical methods.

Statistics estimate the percentage of the time that we would be wrong when we say "There is likely a difference." They do so through probabilities and *confidence intervals* (covered in Chapter 35) for their estimates of differences, using knowledge of the variables and of their distributions. The better their measures correspond to the needs of the statistical techniques, the better their estimates of the probabilities will be. The probabilities are called *p* values (covered in the next chapter).

Both our researchers know that their results are essentially the application of faceless equations from theoretical worlds on local systems where chaos and nonrandomness reign. Risk, probability, potential error, careful wording to accommodate ambiguity, and other specific characteristics of the local situation are all features of the world of statistics. Rather than dark and scary places, they are simply elements of the landscape and of the way things are.

Both the high school principal and the director of public health know two critical pieces of information for this topic: They know that *p* values are the specialized probabilities that are the end-product of statistics, and they also know to interpret them carefully, for reasons discussed in the next chapter.

31. Uncertainty—p Values

End of analysis

Start of results

Thin ice

S tatistics help to tame ambiguity by quantifying it. We decide in advance how large a risk we are willing to take if we are wrong in saying there's a difference, thereby rejecting the null hypothesis. In many situations, people have settled on a 5% risk, hence the frequently seen .05 cutoff for statistical significance. Before accepting someone else's judgment, ask yourself whether that is an acceptable risk for an answer to the question at hand. What if the question were about a meltdown risk to your local nuclear power plant, the one less than a mile down the road? You might want to be a bit surer before accepting the risk of being wrong in your answer. There are other situations where 5% might be too strict because there is little damage from being wrong. Case-by-case judgment can be tedious, but it usually is worth the effort. Willingness to think through these types of issues is a beneficial disposition that can serve a researcher well in many phases of a project.

Here is a traditional method for interpreting a *p* value. When testing a null hypothesis with a resulting *p* value of .04, statisticians would say that results such as these would be due to sampling error only 4 out of 100 times that we might care to rerun the test, if the null hypothesis were true. Next, we might say that the results suggest that the sample probably was derived from different parent populations (i.e., the results are not the same; therefore, they are different).

Notice that statisticians would not conclude anything about the size of the apparent difference based on the *p* value alone because *p* values say nothing about the size of a difference. They only speak to the likelihood that the mathematical difference presented by the group estimates might be due to sampling error instead of being a reflection of the samples coming from different parent populations. Many people make the mistake of thinking that small *p* values imply large differences. They do not. They only imply a greater confidence that the apparent difference seen between groups is a reflection of the samples not coming from the same parent population. Although larger differences can and do create smaller *p* values when nothing else changes, a smaller difference accompanied by a large increase in sample size also can result in a smaller *p* value. Being the result of many calculations, *p* values are compensatory in that they balance several parameters and inputs.

More than occasionally, published reports will contain *p* values that mistakenly are calculated to several decimal places for differences from two groups of about 20 people each. Be wary of misplaced precision in statistics, as it might be an indication of a lack of training in the application of statistics to real-world situations. Remember also that statistics are a form of educated guessing. Presenting statistics calculated to several decimal places suggests a degree of certainty in our models that also seems somewhat misplaced.

Both our researchers are still thinking about these issues. They show *p* values to the four decimal places generated by their software, and both use .05 as a cutoff, yet for different reasons. The high school principal was heard saying, "Why rock the boat?" He rarely had been challenged in the past and has all of his formatting set to accept four decimal places for *p* values. In his view, nobody else cares, so why should he? In many ways, he probably is correct in his thinking about the end result, but he is not correct in his thinking about statistics. He also does not produce high-stakes statistics, meaning that little of real importance changes on the basis of his findings alone. For him, staying with .05 as a risk cut-point seems like a fine standard to use.

The director of public health thinks that the makers of the software must have thought about this issue and, therefore, probably are correct in showing the four decimal points. She has very large samples that probably could support *p* values to four decimal places. She does not realize that the four digits might be there to assist in rounding to two digits for smaller samples. Alternatively, the precision of the output is a definable option in most software. The default number of digits might not have had much thought behind it at all. In any case, her reports also have had four decimals for some years and probably will continue to do so. Some of her statistics, while potentially high-stakes, also are greatly influenced by large samples. She has decided to set her risk cut-point at .01. She believes that the combination of public health importance with large samples justifies setting it conservatively. She could have set an acceptable risk at .001 and still had little problem reaching it for most of her work because her samples are that large. Readers might think that level of risk to be too conservative, missing important findings. Even the determination of acceptable risk can have a political component.

Researchers must be able to defend every decision they make, even when those decisions are to accept the default settings of their software. Researchers might not even be aware that they are making some of these decisions, yet in the end, very few of these decisions are challenged. This lack of challenge can result in statistical work that is not as careful and reflective as it otherwise might be.

32. Expectations—Chi-Square

Flexible

Useful

Friendly

P roportions often are tested for their equivalence (no difference, null hypothesis) using *chi-square* statistics. Although chi-square can take many forms for many related purposes, almost all incarnations ask whether the proportional distribution in one group is equivalent to that of another (or more than one other), without leaning on the normal curve for support. This freedom from the normal curve makes chi-square useful for data that blatantly are not normally distributed. If someone wanted to know whether voting patterns along political party lines differed for school superintendents versus police chiefs, a chi-square would be a good statistical test to use.

Both the high school principal and the director of public health will use chi-square statistics in their work. This class of statistical tests is pretty stable and requires few of the assumptions that underlie many other statistical techniques. In most places where questions about differences can be framed with proportions, chi-square statistics can be used in answers.

Chi-square tests are particularly well suited to discrete systems, where whole numbers are used to count numerators and denominators. The principal uses these tests when asking about differences in extracurricular club involvement by year in school. He looks at the numbers of freshmen, sophomores, juniors, and seniors involved in clubs, and he divides each by the size of its class. But finding a difference does not tell him why it exists. He would use qualitative methods (probably an informal survey of students) where needed.

Although the director of public health often uses chi-square tests, she knows that this class of analytic methods is fairly sensitive to large sample sizes, which she has. She will look first at *p* values but then at the substantive sizes of the difference. For much of her work, trivially sized differences can turn out to be statistically significant because of the very large sample sizes.

33. Importance vs. Difference— Substantive vs. Statistical Significance

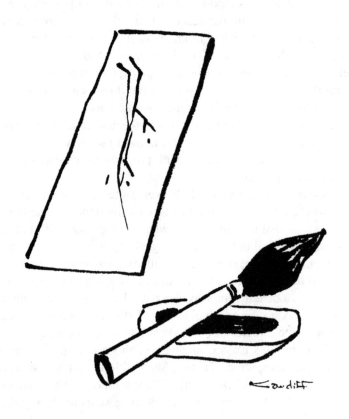

Blindfolded statistics

A flash of insight

People determine worth

J ust because something is *statistically significant* does not mean that it is *substantively* important in addressing the intent of our questions. There might have been too much *statistical power* (i.e., the ability to find a difference, discussed in the next chapter). When there is too much power, differences that are trivially sized and would not make a real difference to anyone can present themselves as statistically significant. Alternatively, a true difference might not look statistically significant due to the test lacking the statistical power needed to find a difference. These issues are weighed and balanced against the resources available to the project and the size of the difference that is likely to be important and able to be found. Lots of guesses and judgment take place at this stage of a research project. Guesses and judgment largely explain (or excuse) why so many reports have seemingly (or outright) contradictory results.

The important issue here for the high school principal and the director of public health is trying to determine the minimum size differences that will be viewed as important to their constituencies, in front of whom they probably will have to defend their results. Both researchers try to appeal either to national guidelines or to community values. The principal presents dropout rates, where having less than 5% qualifies the school for a state bonus. The director discusses rates of immunization trending upward, saving more lives each year.

Both the principal and the director minimize the extent to which they are seen as making arbitrary decisions. The principal knows that his audience wants to see children stay in school, with the bonus of lower local taxes. For the director, keeping more people alive through increased immunizations is a message welcomed by all stakeholders.

34. Strength—Power

Old wisdom

Strength in numbers

Statistics agree

S tatistical power is the ability to detect a true difference, measured on a scale from 0 to 1. Traditionally, research projects set power at 0.80, or an 80% chance of detecting a true difference, if it exists, and rejecting the null hypothesis. Although a few maneuvers can be used to increase statistical power (e.g., looking for a bigger difference), the typical method is to get more subjects (i.e., increase sample size). With too much power, you risk finding trivially small differences. With too little power, you risk not finding real differences that could be important. Because increasing power normally means increasing the size of the sample, it also means that costs can increase quite quickly. Most projects therefore strike a balance and try to find substantively sized differences where resources allow.

The high school principal and the director of public health are in opposite situations for statistical power. The principal has a fairly large percentage of the data that he needs, and he can get at it quite inexpensively. The course sizes in the honors courses, however, are as small as eight students. These group sizes are too small for statistics to yield much of an answer. The principal would need sample sizes that would approach that of an entire class, such as seniors, to support a reasonable amount of power in a complicated model.

The director has huge numbers of subjects, so the potential for too much statistical power is high. Using *p* values alone, she would likely find trivially sized differences for just about any variable she were to use. This issue explains why she needs national guidelines or other standards to help her to determine a valid frame of reference by which to judge her results. Power adds an interesting and an important twist to statistical analyses.

35. Likely Range—Confidence Intervals

Twixt here and there

Most of the time

Safety in disclosure

C onfidence intervals are created to show the likely range of an estimate. Frequently, they are set so that the estimate would fall between the two ends of the range 95% of the time the research might be repeated. A lesser level of certainty would result in a smaller interval but less assurance of it holding true as often. They also are used to show the *likely* range for the size of an effect or for the strength of a relationship.

Every inferential statistic could have a confidence interval wrapped around it. Most should. Confidence intervals show the extent to which estimates could be inaccurate. They can be used for means, correlations, percentages, and all kinds of other statistics. Of the many statistics we see in day-to-day life, perhaps only the national polls seem to give us some measure of the confidence interval, the plus or minus 3% or so.

The high school principal and the director of public health are once more in opposite situations for this issue. The principal does not want to "load down" his report with aspects of statistics that could require him to spend most of his presentations and discussions in technical explanations to audiences who mainly want to focus on why his scores say education is fine, but colleges and industry are bemoaning the poor products emerging from public high school. Although he might appreciate the diversion, he needs to face these questions instead of being pulled into technical discussions.

The director of public health knows that most of her audiences want and expect to see confidence intervals. Her stakeholders have long been exposed to them in both reports and presentations. Even when she has so much statistical power that the confidence intervals are quite small, she will need to show them as standard aspects to her reporting.

36. Association—Correlation

Two people dance
They spin, twirl, more or less together
Do they hear the same music?

Correlations are much like scores in pairs synchronized swimming events. The two people involved can achieve perfect scores by performing identical or exactly opposite movements. Anything in between is not as synchronized. Correlations measure the degree to which values for two measures move in a synchronized manner. The farther away from 0 and closer to 1 (moving together) or to –1 (moving opposite each other), the more synchronized they are. The closer to 0, the more independent they are. Generally, (absolute) values of about 0.2 are considered weak, about 0.5 are considered moderate, and about 0.8 are considered strong. Correlations have two aspects: strength and direction. Most of the time, we want to know both.

Correlation, by itself, does not imply causation. This point is worth remembering well. Correlations assess the strength of mathematical associations. The correlations are not considered to be causal because a different, potentially unmeasured, characteristic might actually be the causal agent. Being human, people state or imply causation more often than they might intend. It is quite easy to slip and state a finding of an association as though it were causal. Many researchers have paid a price in terms of their reputation for implying causation when they only had association.

The high school principal looks at correlations between course grades and many of the other variables. He really wants to understand why the higher-scoring students are not always those who seem the smartest or who do the best on the standardized tests. Shouldn't the brightest students get the best grades, too?

The director of public health likes the more general, overarching, "mile high" look that she gets with correlations. She knows that she will need to build somewhat complex models for many of her questions, but simple correlations help to frame and guide her work.

37. Predictions—Multiple Regression

Name the shape of a cloud

Have others, too

What is it?

P rediction, explanation, and "statistical adjustment" are generally done through one of the many incarnations of *multiple regression* analysis. Multiple regression analysis is the backbone of much of social statistics. Although multiple regression has evolved into several specialized varieties, all forms assess the direct and combined relationships between variation in a single dependent variable's values and variation in (generally) several simultaneously considered independent variables. Questions that range from drug effectiveness to dolphin population sizes are addressed through types of multiple regression analysis. It probably is not in error to suggest that in some student experiment somewhere, the way the wind will blow is being forecast through multiple regression analysis.

Although multiple regression typically uses multiple independent variables (hence its name), special cases of the mathematics behind it underlie some everyday statistics. For example, the mean—the simple arithmetic average—can be found using a specially coded multiple regression analysis. Other common statistics, such as *t* tests, also are special cases of ways that multiple regression can be used. Simply put, multiple regression analysis is a very versatile tool. From testing for differences in group means to testing for differing group slopes (i.e., *trends* of one nature or another rather than *averages*), multiple regression analyses often can be found. The technique is so flexible that many statisticians think in its terms first, because its models have a powerful versatility and have software available to do the math.

Both our researchers will be doing various forms of multiple regression analysis, almost as a matter of course. The high school principal will attempt to predict risk of failure and has multiple measures to use in doing so. He operationalizes failure as a dichotomous variable (i.e., yes or no) and so uses logistic regression. How did he know to use logistic regression analysis? He consulted with his local statistician before beginning. A mistake at this stage could have gone quite far to invalidate his final results, no matter how carefully everything else was done. The fact that he has a dichotomous dependent variable changes the landscape for the questions that can be asked and the way that they can be answered. His statistician will be helping him at several stages of the project as well as when he is writing the report.

The director of public health will be looking at differences in immunization rates for people with various characteristics, using a more traditional multiple regression method (*ordinary least squares*). Her rates are averaged over local areas so that reporting and sharing of the data are not restricted by legitimate privacy concerns and related laws that might pertain to data at the level of individuals. Had she been interested in examining the likelihood that certain beneficiaries would get a pneumonia vaccination (yes/no dependent variable), she would have used logistic regression, too. At this level of detail and potential for identification, several laws restrict what can or cannot be done with the data. These laws are not intended to be lenient with "first-time offenders."

Field-specific legal issues along with substantive knowledge of research unique to the topic of interest are major reasons to hire statisticians with relevant experience. Substantive fields each have their own ways to code data. Even the variables that are deemed important enough to collect, code, and store can differ in fundamental ways. What is not different is the approach to answering questions offered by statistics themselves, a probabilistic view with the major issues accommodated, frequently done through a version of multiple regression analysis.

Some of the more important numbers returned through multiple regression relate to how well the model fits the data. Some may argue that the direction of interest is in how the data fit the model. I maintain that the data are what they are, and it is the model that is being assessed. If the fit is poor but the data are sound, the model must be adjusted, not the data. If the data are not sound, they should not have been used in the first place. Only the director of public health's constituents (not the principal's) are used to hearing results about these "fit" statistics.

With acceptably sound data in place, multiple regression models tell us how effectively the independent variables function as predictors of our dependent variables. Alternatively, the independent variables could be assessed for the extent to which variation in them coincides with variation in the dependent variable's values. It all depends on the question of interest and the statistic chosen to assist with an answer. Multiple regression can tell us the relative amounts by which several variables seem to be associated with a given variable of interest. We can see confidence intervals for them, p values, and other statistics that we are starting to learn more about.

An entirely separate view of multiple regression is as *multiple correlation*. We recently looked at a case of *bivariate correlation* (between two variables). Multiple correlation is between one variable (the dependent variable) and a string of others (the independent variables). A multiple correlation coefficient, though, does not have the option of a positive or negative sign attached. They are all positive for reasons of the mathematics. The strengths of the associations, as measured through the multiple correlation coefficients, are interpreted much the same way as for bivariate correlations.

38. Abundance—Multivariate Analysis

Wind, tide, sails, and heading

Colors fill the bay

An invisible line keeps score

C omplicated questions can require complex models and many covariates to be addressed; this is known as *multivariate analysis*. Additionally, as a society, we are loath (thankfully) to conduct true social experiments. For example, for many questions involving the effect of nutrition and other environmental factors on adult characteristics and choices, society could deliberately separate identical twins at birth and place them in conditions that would inform the differences in question. We do not allow those types of experiments and generally would rebel against anything remotely similar to them. One result of our ethics and morals is that we rarely have conditions between or among groups that can be considered functionally equivalent, a sort of minimum requirement for comparisons between and among groups to make sense. Therefore, lots of values are gathered on lots of characteristics to make the statistical adjustments required to justify a comparison of groups.

Our two researchers are in somewhat different situations for their need to use many variables at once. Each has several variables to consider and balance, and for both, the sea gets a bit rougher when sailing into multivariate waters. The big difference (remember) is that the director of public health has lots of statistical power that can be used to control for the plausible confounds to her results. The high school principal is not as fortunate. He has to try to make his comparisons meaningful through careful planning. For example, he will try to use the same students in different classes or under different conditions when he can, using them as their own frames of reference. The director finds it easier and statistically feasible to control for whatever characteristics her stakeholders find important to use. Whether these characteristics are important to the results is less important to her than is delivering her overall message by way of her results.

39. Differences—t Tests and Analysis of Variance

Departures in means

Separation in groups

The importance?

M uch of statistics looks for group differences in means, usually the arithmetic averages. For differences between two measures for one group (paired) or in one measure for two groups (independent), t tests often are used. Robust and available in a surprisingly wide array of software, t tests are a mainstay of action research (i.e., local research on local problems). These tests are also found in some very high-stakes research. Simple and common does not mean unworthy of use in high-profile areas. Simple can be elegant and powerful when properly applied.

When the number of measures or groups increases, tests for differences in means get a bit more complicated and are referred to as *analysis of variance* (ANOVA). The types of questions to be answered are still structured as before: Are there mean differences in specified measures among selected groups? The change is through an increase in the number of measures or groups being compared simultaneously.

For most scientific reporting available to the general public, when there is more than one measure (except for the paired t test) or there are more than two groups, the ANOVA family of statistical techniques (described next in its component parts) is one of the more common such families in use (along with its close cousin, multiple regression). In a very real sense, a t test is just a special case of ANOVA. Another view would be that ANOVA is simply the extension of t tests to more complex situations. With the main categories of these ANOVA situations being described in the next four sections, the remainder of this chapter will discuss independent and dependent (i.e., paired) t tests.

With simple models, such as t tests, the scores or values are assumed to be justifiably compared without statistical adjustments. This assumption can be stretched a bit, but not a lot. That is one reason why so many different statistics exist: to accommodate various conditions and to take account of whether the groups are independent of each other, which can make a big difference in choosing a statistic or a sampling approach.

For t tests, then, we see two flavors: two measures for one group, and one measure for two groups. The first of these flavors has a few different names: dependent, paired, and one-group t tests. The second is simply called an independent (or two-group) t test. By using balanced research designs with random assignments to starting groups plus some other controls, both types of t tests can be effective statistically and powerful allies politically.

The political aspect is one of the biggest benefits to using t tests. Their recognition by a wide range of audiences and the fact that their familiarity gives them tacit validity make them handy for presentations to diverse audiences. The reason is that people who are used to seeing research results or who occasionally read a layman's version of technical journals will have seen these tests enough to feel comfortable with them and with how they should be interpreted. The interpretation is simple: A statistically significant t test suggests a difference in means between the two groups, all else considered equal—the *ceteris paribus* maneuver.

The high school principal was a psychology undergraduate major and was well exposed to t tests and ANOVA through a series of mandatory courses. He finds that a series of independent and paired tests gives him a better overall picture of a situation than does a choice of one type or the other. That advantage springs from the two tests answering somewhat different types of questions. For independent groups, he wants to know whether a particular difference in the groups is associated with a difference in outcomes. For example, two sections of the same course are taught by two different teachers at the same time of day right next to each other. The sections have randomly assigned students taking the same tests. If these students differed on their course grades, the implication would be that the difference was due to the teachers' influence alone, all else being considered to be equal between the two sections of the course (*ceteris paribus*). Causality might be ascribed through logic and good research design, but discussing strengths of association would be safer. We have seen the extent—weak at best—to which logic is an ally in statistics.

The high school principal essentially wants to know whether students score higher in one section than another. He finds that the average grade of the students in one section is a B and in the other section a B– and that the difference is statistically significant.

Naturally, he talks to the teachers about their expectations and grading procedures—statistics in action.

The director of public health's course of study included biostatistics, which was more concentrated on variations of multiple regression (described earlier) than on ANOVA or *t* tests. For these types of tests, she will get some help. One facet of ANOVA she learned seemed important enough at the time to be worth remembering: The default error terms used by statistical software were almost never the ones needed for her assignments. She could almost count on the results being wrong when she let the software just use its default values. She vowed then and there that a statistician would be close at hand whenever she might need ANOVA, *t* tests excluded. Their reduced simplicity makes them much more user-friendly to researchers and audiences alike.

39.A. ANOVA

Three or more

Even start

Even end?

Straight ANOVA is used when there are more than two groups and no "side" conditions needed to "adjust" the results so that the groups can be justifiably compared. In other words, the groups start off sufficiently equivalent to justify a later comparison on that once-equal trait. Remember, side conditions generally are statistical accommodations of plausible confounds needed to justify a fair comparison across groups when they are not initially equivalent. When possible, it is worth your while to work harder to set up the project so that statistical adjustments are not needed, or at least fewer of them are needed.

One way to think of straight ANOVA, as mentioned earlier, is as a multiple group t test. The reason why several simultaneous t tests shouldn't be run is that the acceptable error rate can slip pretty far fairly fast with multiple comparisons; hence the need for ANOVA.

The high school principal wondered whether there was an effect on nutrition from students being assigned to one of three successive lunch periods, as seen in the mean end-of-year height of the students. Typically, the school cafeteria runs out of some items toward the end of the second lunch period. Students were randomly assigned to lunch periods at the beginning of the year. Their heights were taken, and the mean heights for the lunch periods were the same. The heights were checked again at the end of the year. As long as there was appropriate statistical power, a difference at the end could tentatively be associated with lunch period.

The director of public health generally has "starting differences." The only straight ANOVA that she will be doing is for public health services such as pediatric immunizations, which all children should receive. These types of services also have face validity, which helps with her audiences.

39.B. ANCOVA

Multiple players
Unstack the deck
Now deal

Analysis of covariance (ANCOVA) extends the usefulness of ANOVA by allowing for the use of covariates—what would be confounds if they were not statistically accommodated as covariates. Extending the previous example for the high school principal, what would have had to happen if students assigned to the three lunch periods were of different average heights at the beginning of the year? Each student's beginning-of-year height could be used to adjust for differences in the end-of-year heights. In statistical terms, the beginning-of-year heights would be used to control, or adjust, for differences in the end-of-year heights that could otherwise be attributable to the starting differences. The beginning-of-year height would be used as a covariate. The deck would be unstacked in this way.

Are you starting to hear the language? ANOVA to ANCOVA. ANCOVA uses covariates. It is the covariance of the covariates that controls for unequal starting positions. The language makes sense, but it is still a lot to digest quickly.

In practice, the high school principal's data have too little statistical power to be able to accommodate more than one covariate for just about all his work, except for comparing, say, junior versus senior grades, where he has about 400 students at each grade level. The director of public health will be using covariates for measures such as access to health care specialists. She knows that the number of specialists needed per square mile is different for a city versus a rural area, so she will need to control for population density when she looks at rural versus urban areas. She does not want to confound the issue of access with varying demand, as reflected through population density statistics.

So far, we have used a single dependent variable. That will now change with MANOVA, where multiple simultaneous dependent variables are analyzed.

39.C. MANOVA

More than one measure

More than two groups

Big budget

Multiple analysis of variance (MANOVA) is a somewhat different extension of ANOVA than is ANCOVA. Instead of adding covariates into the mix, it adds other measures that also are targets of our questions. The questions addressed by MANOVA are about multiple dependent variables with (usually) only one grouping variable on the independent variable side, no covariates. To return to the earlier example of height difference across lunch period conditions, the principal could ask: Is there a simultaneous effect on both height and weight across the three lunch periods? Extending the earlier height question, we are now asking about both height and weight, and at the same time. That is the realm of MANOVA, multiple measures on multiple groups.

Because of statistical power, there is not much the high school principal can do with MANOVA. He could do something using his lunch period data, but the questions are not much more than curiosities. He has to keep the lunch periods the way they are due to space, and buying more food is outside the constraints of the current budget. With the data he already has, he might ask that the lunch budgets be shifted some toward the later periods.

The director of public health also does not use MANOVA, but for a different reason. She has found that a single aggregate score for the dependent variable works best for her audiences. She forms that score in one of a number of different ways, such as *factor analysis* (described in Chapter 43). Like many people who work with large data sets and multiple measures, she has found that "simplify, simplify, and simplify" are three useful rules. An example of aggregating scores would be the way she groups pediatric immunizations into a single measure rather than using a separate measure for each. One score is easier to discuss.

39.D. MANCOVA

Multiple measures

Multiple groups and controls

End of the pier

Multiple analysis of covariance (MANCOVA) is the end of the ANOVA pier—and it is a real seatbelt-tightener. It addresses questions about group differences in simultaneous dependent variables in the face of other characteristics that need to be accommodated (covariates). This method addresses questions about, say, simultaneous differences in height and weight among the three lunch period groups of students after controlling for differences in their starting characteristics and nutrition at and away from home. The different starting conditions are different beginning-of-year heights and weights for the three lunch periods. The director of public health could look at the effects of ethnicity on immunization rates and diabetic eye exam rates while controlling for access and average income for the zip code. The foundations for these obelisks of models had better be pretty strong.

In practice, neither the high school principal nor the director of public health is getting anywhere near this particular statistical technique—simultaneous dependent and independent variables, looking for group differences while controlling for starting differences. If MANOVA is walking on pretty thin ice with regard to meeting assumptions and yielding policy-relevant information, then MANCOVA is almost in a class of its own. Think hard about assumptions and data conditions. There are several places in these models where problems occur. All the math and all the computers and software might be in place, but the data typically are not, at least not in a way that they would be needed. Large samples distributed in just such a way are needed to meet the assumptions of the basic model. None of this is lost on the statistical establishment, which might explain why MANCOVA is so rarely seen under real-life conditions. Nonetheless, when done well, MANCOVA can inform policymakers about complex systems in ways that simpler techniques cannot.

40. Differences That Matter— Discriminant Analysis

MANOVA upside down
Or left switched with right
Which differences separate?

*D*iscriminant analysis asks the opposite questions from MANOVA, the flip side of the coin. It takes the group membership as the dependent variable to look for its *predictors* (a term that does not imply causality, which can be the source of a misunderstanding with the press). Discriminant analysis can show the strength of the association between group membership and other characteristics, such as a host of different demographic and environmental characteristics. These strengths of association are shown in a series of coefficients, numbers indicating the relative strength of each variable in choosing group membership. As with other complex models that depend on several measurements, be cautious of results, think hard about the structure and integrity of the data, and look for replication of results in other research for critical issues. As with MANOVA, discriminant analysis is a complex model that requires careful thought for its construction and for the interpretation of its results.

Despite the need for caution, both our researchers will be doing discriminant analyses. The principal wants to know which characteristics might predict students at risk for failure and dropping out, so that policies aimed at effective interventions might be implemented. A discriminant analysis where previous successes and failures are used to "train" the model would start the analysis. He would then enter new students' data into the equations derived from the previous students and dropouts. To the extent that the important characteristics were properly specified and remained stable, the risk of students' dropping out can be predicted.

The director, as we know, is quite interested in rates of pediatric and geriatric immunizations. Before, she was looking for average differences in immunization rates among, say, voting blocs. Now she is looking to see what about those voting blocs might generalize across her state and be helpful in effecting policy changes.

41. Both Sides Loaded—Canonical Covariance Analysis

Multiple measures

Multiple everything

Related how?

C *anonical covariance analysis* relates multiple dependent variables to multiple independent variables all at once, rather like MANCOVA. Instead of looking for mean differences in scores, however, canonical covariance analysis looks at the correlations between all the dependent and all the independent variables. It produces the best weighting of each variable to produce the highest correlation between the two "sides" (dependent and independent variables). Some statisticians will spend their entire careers without needing (or wanting?) to run a canonical covariate analysis, except maybe when they were in graduate school. These models generally are big and filled with many assumptions. In fact, they are usually so loaded with variables and assumptions that there is little agreement on how to test them for sufficient compliance with the underlying assumptions. In addition, very large samples are needed to stabilize the model. This constraint alone places its use beyond the resources of many researchers.

Granted, neither of our researchers is getting near a canonical covariance analysis. Nonetheless, in large data sets that are well constructed and have well-measured variables, these models can inform complex systems in a manner that can be relevant to policy decisions and that cannot be justifiably addressed with simple comparisons. Remember, smaller models often ignore major potential confounds. Again, society thankfully does not permit the types of controlled conditions that would be needed to conduct true experiments with human beings, so we conduct quasiexperiments in natural settings and try to control for whatever systematically occurring conditions we find. If differences were randomly distributed, they would degrade the precision of our estimates but would not necessarily bias them (i.e., predispose them to be falsely high or low). Complex models are needed to make all the necessary adjustments for justifiable comparisons with nonbiased results. Canonical covariance analysis is a flexible way of making adjustments.

42. Nesting—Hierarchical Models

Within the clan

Nuclear families

Have dinner

When situations exist whereby the subjects of an analysis are somehow contained within another group (e.g., patients within doctors, students within specific teachers' courses, or engines within factories), we use a *nested* analysis. Ignoring nesting has caused almost immeasurable damage to egos and reputations, even to science itself. Handling nesting is often a complex issue, but it still should be done.

The issue at hand is that groups tend to be and to act somewhat more like themselves than do randomly picked individuals from across all groups. Sometimes that group effect is something that is the target of interest; sometimes (and more often) it is a plausible confound that needs to be accommodated in the model, one way or another.

If a single most likely place exists where statisticians and policy-makers talk past each other instead of with each other, it is probably when statisticians think they are dealing with a nested model where the statistical effect from the nesting is too big to be ignored. Both terminology and frustration seem to escalate with incredible speed.

The high school principal and the director of public health will try to accommodate their nested issues as best they can, with help from the statisticians. The challenge for the statisticians is that no guide-lines exist to suggest when the issue is small enough to ignore or large enough not to ignore. It is not a game to statisticians, because they might believe that the impact from the nesting is too large to ignore. The problem is more one of language. When statisticians need to support a perspective, the language gets overly technical very quickly. When both parties are not similarly armed, the outcome is predictably and unfortunately one-sided.

43. Cohesion—Factor Analysis

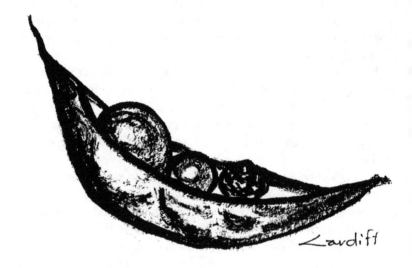

Relatives

Split many ways

Steel-blue eyes join

F *actor analysis* reduces the number of variables in an analysis, and it can do so in a wide variety of ways, each of which means something slightly different from the others. To begin, factor analysis can show the extent to which variables might be viable proxies for each other and might thereby be somewhat redundant. Factor analysis can also show an underlying structure by which items on a survey "hang" together as a group. Many types of factor analysis exist; but regardless of the type, factor analysis seeks to reduce the number of variables by extracting what the original variables have in common into the most parsimonious set that is consistent with how theory suggests the underlying traits should be exhibited. The reason that a large number of types of factor analysis exist is the large number of differing perceptions that people can have about multidimensional problems. Mathematics offers many options, and statisticians seem to have lined up for them in factor analysis. Experience suggests that most statisticians' bookshelves contain at least a couple of books on factor analysis. I am not sure whether the books are there for assistance or as badges of courage from the courses. Either reason probably is good enough.

Factor analysis frequently is used when examining latent traits—those characteristics that cannot be measured directly, such as mathematics knowledge, and that can be inferred only from test performance or other behavioral observations. The structure of the analysis shows how items group in interesting ways. By itself, though, factor analysis can only show a reshuffling of the information content from the original variables. People who know about the variables (or items) need to step in and explain how and where factor analysis makes sense, as well as how and where it does not make sense. This last portion, noting where it does not make sense, can save reputations and careers alike. It has also been the genesis behind long acknowledgment sections in journal articles thanking people for comments on earlier drafts.

Factor analysis is also used to reduce the number of items on surveys. When little new information is gained by adding an item, this analysis highlights that overlap so that one of the items can be removed. Other types of *item analysis* (using correlations between items and descriptive statistics on items) help to decide which of the items should remain and which not. In a way, factor analysis and item analysis have some aspects in common, insofar as they both seek to uncover the most parsimonious subset of items or variables that will sufficiently capture the information contained in a larger set.

One more way that factor analysis is used in complex systems is as a method of aggregating several measures into a single score. The first factor generally is the one used for this purpose. It is the best overall single alignment of the multiple sources of information on the topic. For, say, quality measures in industry, the first factor would represent the best achievement of any of the firms on the overall measure set, automatically adjusting for the measures that are more or less frequently achieved by members of the group.

Both our researchers could use factor analysis, for reasons alluded to previously. The high school principal needs overall measures of academic achievement. Factor analysis will adjust for the tendency of some students to take courses that are graded more strictly than some courses taken by other students. The director of public health needs factor analysis to determine some of the overall health scores given to communities. Factor analysis especially rewards communities that do well on difficult measures compared with measures where everyone does well. The reverse also is true— factor analysis would penalize communities for doing poorly on measures where most are doing well.

Factor analysis almost always requires the services of a statistician. Although some software programs make it easy enough for almost anyone to get an answer, most people want the *correct* one. Remember, the choices are mind-boggling. With this many choices, think of them as options. Apparently, there is quite a bit to think about in getting meaning from the results that the method produces. We do not know what we do not know. It is up to us to wade through the information with a statistician to be able to better understand what is found in research. The method will influence the results. With so many methods, then, there must be an equally large number of potential results that could have been shown instead of the ones that were.

We are quickly approaching a place where statisticians tread lightly. Can we really trust models with so many assumptions that seem to do such magical tricks with everyday data? In its sorting, reducing, and capturing of equivalent information content, factor analysis seems a bit too good to be true, in an old backwoods wisdom type of way.

In some senses, it is. Because factor analysis can be done so many different ways and the results also can be so different, some statisticians are given to pause at the findings from anything but the most routine of methods for this class of statistical techniques. At question is whether this situation is a type of confounding. How much of the results is due to the specifics of the method versus how much is robust and important unto itself? That is what understanding factor analysis is all about—that and reducing the dimensionality (variables, items, etc.).

44. Ordered Events—Path Analysis

A map of events

Relationships shown

The governor signs

P *ath diagrams* (used in *path analysis*) can support causal inferences because of their ability to accommodate different periods in time. They also can be powerful tools for presenting the strengths of relationships for complex issues in a fairly "clean" format. Those are the strengths. Their biggest weakness is that they are models. In social systems, complex relationships can be difficult to capture with valid data. Additionally, path analyses generally incorporate several variables, so somewhat large samples normally are needed. Yet path analytic models have a beautiful look to the eye. They almost scream parsimony. For many years, journals liked them a lot. Many people still do.

Our two researchers are united in their intent to use a path analysis somewhere in their reports. They like the impact, and the analyses split the relationships up into policy-relevant chunks with easily distinguishable groups and traits. Path analytic models show relationships as clearly as road maps show streets and blocks, though some of the models have paths that you would not want to use as a street on a map.

The high school principal has been asked by the school board to build a model of college acceptance on the basis of grades, standardized test scores, level of courses taken, and a few other student characteristics and school experiences. The director of public health wants to model childhood immunization patterns across the various communities in her state on the basis of background characteristics and access conditions. Path analysis should serve them both well.

The researchers keep asking the same general questions but keep going after answers in different ways. That is part of the nature of statistics, just as someone viewing a house from a westerly orientation would describe it differently from someone else viewing the same house from the east. Buyers want to see all views.

45. Digging Deeper—Structural Equation Models

Path analysis

Latent traits, too

More insights

S *tructural equation modeling* is the marriage of path analysis (just seen) with factor analysis (seen just before it). Now, latent traits, such as math ability or propensity to be a compliant patient, can be modeled with all of the benefits of path analysis and factor analysis combined. Structural equation models also can accommodate nested (i.e., hierarchical or grouped) data, but they become difficult to draw on two-dimensional paper at that point. Intellectually, they are appealing models. Graphically, they can be terrific. From a statistical estimation standpoint, they can present a series of problems. As a suggestion, have a large budget for statistical help.

Furthermore, structural equation models have all of the weaknesses of the two combined parent techniques. They rest on many assumptions, some testable, some not. They seem to be susceptible to what would seem like minor changes in their specifications. Yet, when done well, their insights can be quite important for policymakers, high school principals, directors of public health, and others interested in the workings of complex systems.

Structural equation models are pretty far out on the pier. They are not seen too often outside the academic press, so neither of our researchers has the slightest interest in doing them. Their constituencies are not used to them, and quite a bit of audience education would have to take place. This audience education generally is distracting from the main message of presentations. Neither the principal nor the director will be estimating a structural equation model.

The intentions of the principal and the director aside, well-conducted structural equation models can highlight important, policy-relevant findings that are unnoticed by other statistical techniques. Although neither of the researchers will be using structural equations, others who have done so in the past have also advanced our understanding of some complex environments.

46. Fiddling—Modifications and New Techniques

Twisted like so

Transformation

Improvement

S tatisticians are forever refining their models and their techniques. That is how statistics got to where they are today. Keeping current in every aspect of statistics used for a single field of social science research would be more than a full-time job. Yet the improvements to statistics give us better answers to our difficult questions about issues where people deeply care.

Statistics rest on assumptions and mathematics. The ideas and value in statistics rest on how they can help to inform important issues that make differences in people's lives. Remember that statistics need a context. Remember their limitations and where we wish they could give us better answers.

So bring on the new modifications, the better methods for handling nesting and other complex issues, the new techniques. Why? The majority of us, including statisticians, will never have need of the modifications. In general, the more specialized the modification is, the more limited will be its use. Yet, those who do need them will get better answers. Those better answers can then better inform policymakers to make better decisions. Statistics is in business to make the most educated guesses it can. If the world is indeed influenced in some small manner by statistics, those statistics should be as informative and accurate as possible. To the extent that modifications aid that process, more techniques are better than fewer.

As you start down your new path of understanding, remember to be skeptical and keep turning over the rocks to find dirt. Statistics are probabilistic answers that are based on assumptions. First ask: Does the answer make sense? That question is not an absolutely reliable truth detector, but it has been shown to be an excellent overall guide to sniffing out bad statistics and recognizing (and now better understanding) the good ones.

47. Epilogue

Long reflect
On a hurtful act
Where little is gained
And you cannot take it back

A s we emerge from the world of statistics, we see that order is brought to chaos at a cost. That cost is certainty. Incredibly, the benefit from the seeming loss is enormous. Without needing to know how to do statistics, we are now more aware of why and how they can be wrong, as we suspected they might sometimes be. As important, we are also aware of why and how statistics are more often correct. To transform this new awareness into an intuitive understanding of patterns and trends requires study, practice, and reflective thought.

Once we internalize the uncertainty, we see reports with divisive conclusions written in absolutist terms as somehow unfair. We want to see the weaknesses of the research made clear. Our ethical standards seem to keep pace with our increasing knowledge of statistics. We need to be on stronger ground to be assured or before making statements that could be hurtful to others.

Later, we become more aware of ethics in reporting as a broader research issue. We might question the specifics of methodologies, statistical techniques, or assumptions of the data. We think of how patterns and trends in the handling of the various issues that can arise can lead to an almost too convenient set of results.

Eventually, we cannot help but see when findings seem overstated or when assertions do not seem supported by the reported statistics. At that moment, we reach a fork in the path of understanding with statistics. To truly understand the extent to which a report is faithfully representing the statistical results, knowing the math is essential. Granted, with good intuition and without the math, we can be aware of when we think there could be a problem. We might even be correct fairly often. Only with the math, however, are we capable of finding out for ourselves the more likely answer and having a deeper understanding of the strengths and limitations of those results.

At this point, some people will take the time to learn the math. Others will be content with knowing more about the topics being discussed than they did before. Both paths continue for quite a spell beyond the fork itself.

For the people who continue down the path with the math and equations, their understanding moves to an entirely new level. Their intuitions will be guided by a more honed knowledge of the issues that matter to a model and to the statistics that result.

All of which brings us back to the simple fact that the sophisticated knowledge of a topic is a better predictive foundation than is randomness. Remember the coins, dice, and cards that helped forge a bond between mathematics and economics (i.e., gambling). This merger played an important role in establishing an expectation for the underlying mathematics to be dependable. Money was at stake.

Over time, the questions became more complex and came to concern social or other difficult-to-count issues. The number of unknowns in equations increased at a fantastic rate. The push toward parsimony was (and still is) under tension from the desire to explain more variation and predict more accurately. Models grew, changed, and adapted to changes in theories and advances in the underlying mathematics.

To see the tao of statistics in action, watch the face of a statistician who is successful at weaving together threads of associations or uncovering differences in new and important ways. Knowledge is born. New questions are asked. For many, a flash of insight brings a smile across their face as they see the next steps along their path and wonder what might lie ahead.

With assumptions piled high

Armed with knowledge and intuition

Although mistakes can be made

Don't wager against a statistician

About the Author

Dana K. Keller, PhD, has explored Taoist and other Eastern philosophies for more than three decades, with journeying to and teaching in China and Tibet as parts of that exploration. He embraces two very different worlds, the West with its scientific approach to knowledge and the East with its more balanced approach to experiencing life. In *The Tao of Statistics*, he presents a way that the two worlds can coexist harmoniously and benefit from each other. His PhD is in measurement, statistics, and program evaluation. After supervising the research component for more than 100 doctoral dissertations and serving as a charter member of Delaware's Panel of Evaluation Experts, he joined the Delmarva Foundation as its chief statistician. During his 7 years serving in that capacity, the Centers for Medicare & Medicaid Services named him as a recommended national resource for the nation's managed care organizations for resolving complex sampling and research issues. His almost unique ability to explain statistical and methodological constructs in everyday language has resulted in his being requested frequently as a presenter and technical expert panel member in the public health community. Now president of Halcyon Research, Inc., he continues to bring his ability to explain statistical concepts simply to an ever-widening audience. He can be reached at dana@halcyonresearch.com.